Successful Intellige

Grow

Analytical Mindset

Master Critical Thinking, Reason with Logic, Improve Problem-Solving Skills, and Actualize Your Inner Idealist for Excellence

Devi Sunny

GRAB YOUR FREE GIFT BOOK

MBTI enumerates 16 types of people in the world. Each of us is endowed with different talents, which prove to be the innate strength of our personality. To understand the deeper psychology of your personality type, unique cognitive functions, and integrated personality growth path, you can scan the QR code below or visit www.clearcareer.in for a free download –

"Your Personality Strength Report"

Successful Intelligence Series

1) Book 1 **Grow Practical Mindset**

Fearless Empathy Series

1) Book 1 **Set Smart Boundaries**
2) Book 2 **Master Mindful No**
3) Book 3 **Conquer Key Conflicts**
4) Book 4 **Build Emotional Resilience**
5) Book 5 **Develop Vital Connections**

Clear Career Inclusive Series

1) Book 1 **Raising Your Rare Personality**
2) Book 2 **Upgrade as Futuristic Empaths**
3) Book 3 **Onboard as Inclusive Leaders**

Contents

About the Book 5

Introduction 7

1. Choosing Analytical Mindset 11

2. Psychology of Thinking Fast 23

3. Dangers of Cognitive Bias 33

4. Benefits of Thinking Slow 46

5. Successful Problem Solving 57

6. Strategies for Right Decisions 69

Conclusion 82

About the Author 84

May I ask for a Review 85

Preview of Previous Books 86

Acknowledgement 104

References 105

About the Book

How does one read or perceive situations? It's fascinating how people look at the same thing from different angles. What causes these differences in thinking? It all boils down to the values they prioritize.

Some folks, often called "Feelers," give a lot of importance to their emotions. Emotions can lead to quick decisions as quickly as clouds pass swiftly in the sky. But here's the thing: Are these speedy decisions made after considering the whole picture or after sufficient time to think things through? Feelers might miss important details and facts because they're in a hurry, which could result in making mistakes or wasting time.

However, using logical thinking doesn't mean ignoring emotions. It means figuring out why we feel a certain way. Understanding the reasons behind our feelings helps us think more carefully and fairly.

When we're trying to grow or improve, problems often show up. So, how do we solve them? Can we avoid them altogether? Making good decisions can help us sidestep specific issues, especially the ones we can't control. So, what are the best ways to solve problems? How do people make intelligent choices?

This book is part of a series about being smart in different ways. It discusses why logical reasoning and critical thinking are crucial when dealing with problems. It shows that these skills can be learned and aims to help readers develop an analytical mindset, which is vital for well-rounded thinking and personal growth.

Introduction

"Simple can be harder than complex: you have to work hard to get your thinking clean to make it simple. But it's worth it in the end because once you get there, you can move mountains." - Steve Jobs.

Have you ever envisioned a world pervading only with happiness? Such a world would signify an absence of problems or deficiencies. It seems like a static existence. If humans possessed everything they needed, what would drive them to strive for more? Every invention typically emerges from a quest to improve life, as necessity often catalyzes innovation. Problems are guideposts, directing us toward solutions that lead to sustainable happiness.

Consider the problems faced by past generations; have all of them been solved? Take, for instance, the necessity for long-distance communication. When resources were confined to specific locations, individuals often had to relocate, yet yearned to connect with those they left behind. From the advent of fax machines to telephone calls to the internet, electricity paved the way for numerous inventions that brought joy to people's lives. Is perpetual happiness the goal these inventions pursue? Yes.

Yuval Noah Harari delves into humanity's persistent quest to overcome mortality, suggesting that scientists view death as a technical challenge with a potential solution. As a historian and philosopher deeply engaged on the subject of death, Harari co-explored with Daniel Kahneman the reality of death becoming a choice in the future, alluding to advancements in direct brain-computer interfaces as the critical catalyst for this transformative possibility.

"In the pursuit of health, happiness, and power, humans might gradually alter one characteristic after another until they cease to be recognizably human." - **Yuval Noah Harari.**

Science has discovered painless ways to live life, as seen by the invention of sedatives and painkillers brought on by individuals skilled in chemistry and pharmacy engaging in extensive thinking and rigorous experimentation. After numerous trials and errors, effective solutions emerged; businesses invested in these solutions, making them accessible to the broader populace. Take childbirth, for example; countless individuals devoted themselves tirelessly, diligently seeking sustainable remedies in the quest for experiencing "true happiness," such as pain-free childbirth.

Genuine happiness stems from effort, not merely intent; from action, not just imagination.

"To attain real happiness, humans need to slow down the pursuit of pleasant sensations, not accelerate it." - **Yuval Noah Harari.**

Psychologist Robert Steinberg, who proposed the Triarchic Theory of Intelligence, discusses passionate love in his triangular model of love—an emotion quick to develop and fade. Those inclined toward fostering love, peace, and a happy environment—individuals who prioritize feelings (introverted feeling - Valuing/Authenticity and extroverted feeling - Harmonizing/Connecting)—must understand that commitment, thoughtful consideration, and hard work are essential to manifest their desires. Perpetual happiness is not sustained solely by reveling in good times; the importance of nurturing a practical and analytical mindset lies in purposeful actions.

What constitutes the right action? It stems from correct thinking. Studying thoughts has been a continuous pursuit from ancient philosophers to modern psychologists.

"Be a free thinker and don't accept everything you hear as truth. Be critical and evaluate what you believe in." - **Aristotle.**

Thinking requires effort. Sound thinking demands consideration of various facets, resulting in simple and immediate decisions. Daniel Kahneman, in his book "Thinking Fast and Slow," introduces System 1 and System 2 thinking, highlighting that individuals in a happy state tend to be more creative and intuitive yet less vigilant and more susceptible to logical errors. This underscores the necessity for creative, intuitive, and feeling individuals to refine their thinking abilities.

"Intelligence is not only the ability to reason; it is also the ability to find relevant material in memory and to deploy attention when needed." - **Daniel Kahneman.**

In subsequent chapters, this discussion will delve into the 5W and 1H of cultivating an Analytical Mindset, exploring logical reasoning, critical thinking, effective problem-solving, and improved decision-making. Logical thinking involves the consideration of facts and trends. In contrast, critical thinking aims to mitigate biases and facilitates making informed decisions rather than relying solely on emotional responses and intuitive insights. As the second installment in the "Successful Intelligence" series, this book aims to assist readers in developing their analytical intelligence.

1. Choosing Analytical Mindset

"We focus on our goal, anchor on our plan, and neglect relevant base rates, exposing ourselves to the planning fallacy. We focus on what we want to do and can do, neglecting the plans and skills of others. Both in explaining the past and in predicting the future, we focus on the causal role of skill and neglect the role of luck. We are, therefore, prone to an illusion of control. We focus on what we know and neglect what we do not know, which makes us overly confident in our beliefs."- Daniel Kahneman.

Ratner's ascent within the family business during the 1980s resulted in a remarkable expansion of a chain of jewelers. This transformation was marked by unconventional marketing strategies, such as eye-catching fluorescent orange posters promoting affordable pricing for a wide range of products. Despite the industry perceiving Ratner's approach as "tacky," it resonated strongly with the public, contributing to the success of stores like Ratners, H. Samuel, Ernest Jones, Leslie Davis, Watches of Switzerland, and a network of over 1,000 shops in the United States, including Kay Jewellers.

However, the turning point came when Ratner delivered a speech at the Institute of Directors conference at the Royal

Albert Hall on April 23, 1991. During this address, he made disparaging remarks about some of the company's products, stating that items like cut-glass sherry decanters were sold at low prices because they were "total crap."

The fallout from Ratner's speech was immediate and severe. The public backlash was immense, leading to a significant decline in customers visiting Ratner shops. This consumer reaction resulted in the staggering devaluation of the Ratner group by approximately £500 million, nearly causing the group's collapse. Consequently, the group rebranded itself as the Signet Group in September 1993. The incident is often called "doing a Ratner," which refers to making a disastrous public statement.

Why did he choose to use the phrase "Total Crap"? Did he genuinely intend to convey the same meaning about his products? Was he attempting to inject humor or creativity into his communication? Unfortunately, his attempt at light-heartedness resulted in unintended consequences, given his influential role and its detrimental impact on his company.

Suppose he found himself among a group of investors. Would he employ similar language? Would his mindset prioritize strategies conducive to financial gain? Would he risk using such words in that scenario? Considering the context and the financial implications, he would unlikely consciously choose such language that might jeopardize his credibility and potential investment prospects.

You will see it's about choosing the right mindset- A thinking system Ratner should have chosen differently irrespective of situations.

Analytical Mindset

Developing an analytical mindset involves cultivating the ability to pause and refrain from rushing to conclusions or communicating based solely on momentary feelings, regardless of whether one is under pressure or experiencing happiness. It entails dedicating time to comprehend situations thoroughly, assimilate facts and trends, and adopt a more disciplined and patient approach to handling any situation. An analytical mindset means not allowing moods or habitual response patterns to dictate actions but taking deliberate steps to understand, assess, and respond based on reasoned analysis and a comprehensive understanding of the circumstances.

"Moneyball," the 2011 biographical film directed by Bennett Miller, vividly portrays the revolutionary shift in baseball's conventional wisdom. Focused on the Oakland Athletics' 2002 season, the movie centers on Billy Beane (Brad Pitt), the team's general manager. Faced with financial constraints after losing key players, Beane, alongside assistant GM Peter Brand (Jonah Hill), defies traditional scouting by implementing sabermetrics, an analytical approach to player evaluation.

The film underscores the clash between Beane's analytical vision and the entrenched beliefs of baseball's traditionalists. As Beane navigates challenges, his innovative strategy of focusing on undervalued talent through statistical analysis challenges the established norms of the sport.

Moreover, it highlights the pivotal moment when the Boston Red Sox owner, John W. Henry, recognizes sabermetrics as the future of baseball after witnessing its impact through Beane's methods. This recognition prompted the Red Sox to adopt a similar data-driven approach, leading to their subsequent triumph in the World Series. It solidifies the influence and importance of an analytical mindset in the game.

Triarchic Theory of Intelligence

Sternberg's Triarchic Theory of Intelligence transforms traditional views by emphasizing the vital role of adapting to real-world changes. The theory unfolds through three interconnected components, each delving into the intricate mental processes governing problem-solving and decision-making.

1. **Componential (Analytical):** Think of this as the brain's strategic planner. It excels at solving academic problems, partnering seamlessly with the analytical thinker—a true expert at breaking down complex issues.
2. **Experiential (Creative):** The creative side kicks in when faced with new situations or needing a fresh perspective. Picture the brain's creative

thinker collaborating with the doer gear to execute plans innovatively.
3. **Practical (Contextual):** Enter the street-smart part. The brain adapts to changes, shapes the environment to fit needs, and selects the best goal setting. This gear works with the doer gear and information scouts to navigate real-life situations effectively.

Imagine all these gears working in harmony within your brain. The planner sets the strategy, the doer executes, the creative thinker introduces fresh ideas, and the practical gear ensures it aligns with the real world. Sternberg recognizes the diversity of human intelligence, suggesting that individuals may possess a distinctive blend of this intelligence.

Analytical Intelligence

The American Psychological Association (APA) Dictionary of Psychology defines analytical intelligence within the framework of the triarchic theory of intelligence as comprising the skills typically assessed by traditional intelligence tests. These skills encompass analysis, comparison, evaluation, critique, and judgment abilities. Analytical intelligence permeates daily life by enhancing communication with data-driven precision, aiding problem-solving with result-oriented strategies, fostering compromise through critical thinking, improving business efficiency via data analysis, and offering invaluable insights for informed decision-making in research endeavors.

Applying Analytical Intelligence

In many cultures, arranged marriages remain a prevalent tradition where families are responsible for finding suitable alliances and arranging marriages. In the past, the bride and groom often met for the first time on their wedding day, raising questions about the rationale behind this seemingly 'blind' tradition.

However, there's an interesting correlation between the prevalence of arranged marriages and lower divorce rates, notably seen in countries like India. Marriage, deeply rooted in societal and cultural norms, is viewed as an institution nurtured by these values. In arranged marriages, elders or involved parties often assess the compatibility of couples across various aspects. After discussing terms and conditions that aim to ensure the security of their relationship, the couple's consent is also sought. This thorough process suggests that more thought and deliberation might go into these relationships before formalizing them, involving the couple and others in the decision-making process.

The longevity of arranged marriage systems across generations implies that they have evolved and assimilated valuable learnings. It isn't necessarily about advocating for arranged marriages over other forms, whether self-arranged or family-arranged, but rather about considering the decision from an analytical perspective rather than solely an emotional one.

"It is a capital mistake to theorize before one has data. Insensibly one begins to twist facts to suit theories, instead of theories to suit facts." - **Arthur Conan Doyle**.

Beyond personal considerations, an analytical mindset also holds relevance in professional spheres. It assists in making well-informed decisions and discovering the best solutions by thoroughly evaluating all relevant factors.

According to Dave Cornell (PhD), the 12 professions below showcase the application of analytical intelligence across diverse domains. They demonstrate the multifaceted nature of analytical thinking, emphasizing its significance across various fields.

Financial Analyst: Requires advanced statistical analysis skills, handling complex data for investment advice, managing substantial funds, and avoiding costly mistakes.

Pharmaceutical Chemist: Involves extensive knowledge of biomedical analysis, pharmacology, and chemistry for drug development, demanding high analytical and scientific expertise.

Academic Writer: Demands thorough research, critical evaluation of information, and persuasive arguments in essays or opinion pieces, influencing policies and public opinions.

Food Critic: Analyzes taste nuances, ingredients, and overall dining experience, demanding acute sensory perception akin to processing complex data.

Strategic Planner: This entails predicting future trends, understanding multifaceted factors, and decision-making based on extensive experience and broad data analysis.

Editor: Requires scrutinizing writing quality, character development, and storyline coherence, employing multiple analytical aspects in evaluating literary work.

Director: Involves scene assessment, dialogue coherence, and post-production evaluation, employing analytical judgment to create compelling film storytelling.

Doctor: Necessitates comprehensive analysis, diagnosis, and evaluation of multifaceted medical conditions, showcasing high-level analytical prowess.

Mechanical Engineer: Involves designing complex systems, assessing efficiency, and meeting various regulations, utilizing analytical intelligence extensively.

Supply Chain Logistics Manager: Requires managing intricate global supply chains, analyzing multiple factors, and overcoming diverse challenges with analytical thinking.

Futurist: Predicts future trends through historical and current trend analysis, requiring synthesizing diverse data for insightful predictions.

Homicide Detective: Utilizes analytical skills in crime scene evaluation, distinguishing crucial evidence, and solving complex murder mysteries.

Analytical Intelligence for Business Success

According to Indeed.com, Analytical intelligence finds broad application in various facets of daily life, playing a pivotal role in decision-making and problem-solving across multiple domains. Delving deeper into its influence, here are the specific areas where this form of intelligence is instrumental in shaping outcomes and driving advancements

1. **Communication:** An analytical approach emphasizes data over emotions, allowing for precise, fact-based communication. This aids in supporting statements with specific metrics, enhancing brand campaigns, and decision-making.
2. **Problem-solving:** It involves defining, analyzing, and solving problems by identifying causes and developing effective solutions. Analytical intelligence helps quantify results and create comprehensive reports for effective communication.
3. **Critical thinking:** This skill involves analyzing observations, evidence, and arguments to make informed judgments. It aids in breaking down complex problems, understanding various perspectives, and facilitating compromise.

4. **Data analysis:** Utilizing analytical intelligence, individuals gather, inspect, and interpret data to detect patterns and make informed decisions. It improves efficiency in supply chains, production, and marketing techniques.
5. **Research:** Analytical research involves critical thinking skills to examine relationships between variables. This supports decision-making by understanding business operations, processes, and customers, contributing significantly to business growth.

Analytical intelligence is essential for business growth in multiple ways:
1. **Identifying product demands:** It aids in collecting and interpreting data to predict customer needs, enabling companies to prepare for increased demand and prevent shortages.
2. **Satisfying customers:** Analytical intelligence helps in understanding demographics, analyzing competition, and collecting customer data, enabling businesses to cater to customer needs and ensure continued growth.
3. **Improving company operations:** Analytical thinking identifies areas for improvement in business procedures by analyzing, problem-solving, and optimizing processes. This approach increases

efficiency, enhances product quality, and improves user experience.

Analytical intelligence is crucial for effective decision-making, problem-solving, and optimizing processes across various domains, leading to improved business outcomes and growth.

Personality Types and Cognitive Functions

In the first book of this series, we explored personality types, particularly the widely utilized MBTI Personality type tests featuring 16 distinct personalities. Counseling psychologists commonly employ these tests for personality assessments. The cognitive functions associated with each personality define their mental abilities, encompassing learning, thinking, reasoning, memory recall, problem-solving, decision-making, and attention. Primary functions, such as leading, assisting, supporting, relief, and ambition, influence our natural responses. Shadow functions come into play when dominant primary functions encounter challenges.

There exist eight cognitive functions:
- Introverted Intuition (Ni)
- Extroverted Intuition (Ne)
- Introverted Sensing (Si)
- Extroverted Sensing (Se)
- Introverted Thinking (Ti)
- Extroverted Thinking (Te)
- Introverted Feeling (Fi)

- Extroverted Feeling (Fe)

Each of the 16 personality types uniquely combines these cognitive functions, profoundly shaping one's thoughts and actions.For a more profound grasp of the psychology shaping your personality, delve into the unique cognitive functions that further delineate it by downloading.

"Your Personality Strength Report."
https://clearcareerinclusive.ck.page/uniquegift

Developing an analytical mindset could be beneficial if you are inclined towards feeling. Conversely, for thinkers, cultivating a creative mindset might be advantageous. However, both feelers and thinkers often succeed when working with a practical mindset.

Mindset Practice

Identify your personality type and cognitive functions to ascertain your primary mindset: Practical, Creative, or Analytical.

2. Psychology of Thinking Fast

"Recall that people like to do what most people think it is right to do; recall too that people like to do what most people actually do. People have a strong tendency to go along with the status quo or default option."- Richard H. Thaler.

In the late 1990s, Internet-driven startups emerged, aspiring to revolutionize industries. Many, like Pets.com, Boo.com, and Beenz, failed despite their innovative concepts, seemingly ahead of their time. In contrast, Amazon stood out due to its financial approach rather than its products. The pivotal factor was the Cash Conversion Cycle, measuring a company's speed in receiving payment for sold goods. Amazon's triumph lies in its efficient cycle: swift consumer payment, minimal inventory, and extended supplier credit. This resulted in a negative cycle, where Amazon received payment before covering the costs of goods sold, fueling its rapid expansion. Amazon's adept handling of this cycle exemplifies its profound impact on a business's cash flow potential. It showcases how efficiently managing the time between purchasing inventory, selling products, and receiving payments can dictate a company's success. The Cash Conversion Cycle, often overlooked, is a potent metric in assessing and leveraging a business's cash flow, as evidenced by Amazon's exponential growth through its adept manipulation of this financial strategy.

The winners showcased adaptability, innovative strategies, and a focus on practicality and profitability. At the same time, the losers often needed more planning and funding support and underestimated the complexities of the competitive market.

"The combination of loss aversion with mindless choosing implies that if an option is designated as the "default," it will attract a large market share. Default options thus act as powerful nudges."- **Richard H. Thaler.**

What exactly is "brain power"? What distinguishes individuals in this process? Do winners think more or slower? What's the concept of thinking fast? Why do people opt for fast thinking? Understanding the psychology behind these questions prompts us to reconsider our perceptions.

System 1 and System 2 Thinking

In his book "Thinking, Fast and Slow," Daniel Kahneman delves into the intricacies of two cognitive systems: System 1 and System 2. System 1 thinkers generally exhibit ease and a positive mood, while System 2 necessitates deeper thinking rooted in prior experiences or anticipated threats—factors often overlooked by System 1 thinkers. Good moods and emotions often propel System One thinking, while System 2 is triggered by a vigilant mind, particularly one in a negative mood.

"A happy mood loosens the control of System 2 over performance: when in a good mood, people become more intuitive and more creative but also less vigilant and more prone to logical errors. Here again, as in the mere exposure effect, the connection makes biological sense. A good mood is a signal that things are generally going well, the environment is safe, and it is all right to let one's guard down. A bad mood indicates that things are not going very well, there may be a threat, and vigilance is required."- **Daniel Kahneman.**

Furthermore, Kahneman emphasizes that Intuition thrives when wielded by experts in a specific domain. He illustrates this with examples such as a chess master's adept prediction upon scanning the board or an experienced fireman's ability to discern threats without clear indicators. However, Intuition can only falter in complex environments if one possesses expertise in the field.

Creative individuals often operate in a positive mood, fostering the flourishing of their ideas. Yet, the sustainability of this good mood remains in question as life's challenges serve to enhance our cognitive capabilities. Those who think swiftly might potentially align themselves as happiness advocates, drawing from its positive influence on cognitive processes. Idealists tend to align more with System 1 thinking. This cognitive system often characterizes individuals with a positive outlook, relying on intuitive and quick decision-making processes. Idealists may lean towards this mode of thinking due to their emphasis on positivity, optimism, and intuitive approaches in navigating various situations. However, it's important to note that this is a generalization, and individuals may exhibit a blend of System 1 and System 2 thinking traits based on diverse circumstances and personal characteristics.

"Intuitive versus analytical? That's a foolish choice. It's foolish, just like trying to choose between being realistic or idealistic. You need both in life." - **Mae Jemison.**

Primary Cognitive Thinking Pattern of Idealists

Idealists, identified as NFs—comprising INFPs, INFJs, ENFPs, and ENFJs—are characterized by intuitive and feeling-oriented traits. Their cognitive functions define their mental capabilities, encompassing learning, thinking, reasoning, remembering, problem-solving, decision-making, and attention. Each personality possesses primary functions—leading (strength), assisting-supporting (growth), relief (Slightly developed), and ambition (Least developed).

INFPs exhibit Introverted Feeling (Fi) as their strength, while ENFPs' growth relies on developing Fi. Introverted Feeling is the core of authenticity, dealing with morals and personal beliefs. It nurtures a deep appreciation for existence, life, and values. A developed Fi continually drives individuals to evaluate their beliefs, enhancing their understanding.

ENFPs thrive in Extraverted Intuition (Ne), while INFPs' growth stems from the development of Ne. Extraverted Intuition fuels creativity and the exploration of multiple possibilities by connecting events and people's actions. This function's optimistic and creative nature empowers them to excel in roles driven by creativity, adapting swiftly to new situations.

INFJs possess Introverted Intuition (Ni) as their strength, while ENFJs' growth is rooted in developing Ni. Introverted Intuition grants an acute awareness of actions' interconnectedness and future consequences. It allows for foreseeing potential outcomes and patterns, often leading to 'Aha' moments. INFJs excel in drawing patterns from information, cautioning others about impending situations, and persuading them to alter their course of action.

ENFJs excel in Extroverted Feeling (Fe), whereas INFJs' growth lies in developing Fe. Extroverted feelings propel individuals to strive actively for harmony within their social circles, often prioritizing others' happiness, sometimes at their own expense, to feel fulfilled.

The Shadow Cognitive Thinking Patterns of Idealists involve distorted or counterproductive use of Ni, Ne, Fi, and Fe, indicating that system one thinking is ineffective for idealists under stress.

Integrated Personality Development for Idealists

Idealists are creative individuals who often operate at their best in relaxed environments. They exhibit heightened sensitivity, which is challenged in competitive settings based on their cognitive functions. They naturally prefer fast thinking and need to develop slower thinking. For Idealists, less developed functions often include Ti and Te—Introverted Thinking and Extroverted Thinking, respectively.

Introverted thinking involves contemplating actions based on logical reasoning, aiding in thoughtful decision-making and judgments. It facilitates systematic analysis and action plans aligned with personal values. Extroverted thinking brings structure to planning, organizing, and decision-making based on rationality and data. It optimizes processes for efficiency and sustainable, long-term vision.

Dr. Dario Nardi proposes the 'The Magic Diamond' concept, emphasizing integrated judgment for wiser decision-making and keener perception by employing opposing functions. Integrated judgment involves thinking (Ti & Te) and feeling functions for decision-making and organization based on objective, logical, and impersonal criteria, considering long-term impacts. This integration leads to a more comprehensive and balanced decision-making process for idealists.

"Wholeness is not achieved by cutting off a portion of one's being, but by integration of the contraries. - **Carl Jung.**

Power of Fast and Slow Thinking in Decision-Making

In a compelling excerpt from *Scientific American*, psychologist Daniel Kahneman delves into the intricacies of the human mind's dual systems: System 1 and System 2. These systems, outlined in Kahneman's book "Thinking, Fast, and Slow," shed light on the distinct mechanisms—fast, automatic thinking (System 1) and slow, deliberate cognition (System 2)—that significantly shape perception and decision-making processes.

System 1 operates swiftly and automatically, effortlessly processing tasks like recognizing emotions from facial expressions (such as perceiving anger in a woman's face), answering simple arithmetic, and executing routine actions like driving a car on an empty road. In contrast, System 2 engages in effortful mental activities demanding attention and concentration, like solving complex computations or critically analyzing situations.

Kahneman presents vivid examples illustrating the functioning of these systems: from intuitive responses like instantly knowing 2 + 2 = 4 to the mental strain involved in solving multiplication problems like 17 × 24. He emphasizes how System 1's automaticity, while efficient, can lead to biases and cognitive illusions, underscoring the necessity for System 2 to intervene in instances where errors or conflicts arise.

The narrative characterizes these systems as a psychodrama within the mind, portraying their strengths, limitations, and interactions. Through this exploration, Kahneman aims to offer insights into decision-making processes, urging readers to understand these systems' roles in enhancing personal and societal problem-solving. Author Martin G. Moore shares insights on decisive leadership, emphasizing rapid, effective decision-making over seeking unanimous approval. His eight elements for optimal decision-making advocate considering diverse viewpoints, proximity to the issue, addressing root causes, accountability, holistic impact, short and long-term value, effective communication, and timeliness. This approach, outlined in "How to Make Great Decisions, Quickly" on HBR, enhances leadership and team outcomes. Martin G. Moore's principles primarily endorse slow (System 2) thinking, recognizing System 1's automaticity, bias-proneness, lack of logic, and tendency toward errors and involuntary actions.

"The moral is significant: when System 2 is otherwise engaged, we will believe almost anything. System 1 is gullible and biased to believe, System 2 is in charge of doubting and unbelieving, but System 2 is sometimes busy, and often lazy. Indeed, there is evidence that people are more likely to be influenced by empty persuasive messages, such as commercials, when they are tired and depleted."- **Daniel Kahneman.**

Fast Thinking and Purchase Decisions

Have you noticed advertisers employing psychology in advertisements? Their typical goal is to evoke emotions, trigger insecurities, leverage our fear of uncertainty, or offer solutions to urgent issues. Psychology significantly influences all individual decisions. The idea of sales psychology centers on the notion that emotions and deeply held values compel people's actions. An in-depth comprehension of these values, beliefs, and underlying factors driving sales decisions can aid sales professionals in generating more sales. According to Gerald Zaltman, a professor at Harvard Business School, 95% of buying choices are rooted in unconscious emotional ties.

Leveraging Fast Thinking for Behavioral Change

Fast thinking, or System 1, operates automatically, swiftly, and without conscious control. It facilitates quick responses, saving time in routine tasks and enabling immediate adaptation to familiar or survival situations. Its efficiency lies in the ability to make rapid decisions, crucial for swift actions in scenarios where immediate reactions are necessary, ensuring adaptability and quick adjustments to various circumstances without requiring deliberate effort.

Richard H. Thaler, in his book 'Nudge: Improving Decisions About Health, Wealth, and Happiness,' highlights the inevitability of influencing people's choices. He emphasizes guiding individuals toward desirable behavior using subtle cues and nudges, avoiding any implication of superiority over established norms.

"A nudge, as we will use the term, is any aspect of the choice architecture that alters people's behavior in a predictable way without forbidding any options or significantly changing their economic incentives."- **Richard H. Thaler.**

Thaler exemplifies this concept with instances such as housefly images etched into urinals at Schiphol Airport, improving men's accuracy in aiming by capturing their attention. He also discusses the role of a "choice architect," responsible for shaping decision-making environments.

Furthermore, Thaler stresses the potential of inertia to bring about positive changes by understanding and leveraging human behavioral tendencies without imposing decisions. These insights emphasize the essence of nudging, offering ways to create environments that encourage better decision-making while preserving individuals' freedom of choice.

Mindset Practice

Recall the instances where your decisions have been successful and unsuccessful, and categorize them into System 1 and System 2 thinking.

3. Dangers of Cognitive Bias

"Freethinkers are those who are willing to use their minds without prejudice and without fearing to understand things that clash with their own customs, privileges, or beliefs. This state of mind is not common, but it is essential for right thinking." - Leo Tolstoy.

The unsuccessful 1961 Bay of Pigs invasion was a military operation orchestrated by Cuban exiles with U.S. support, aiming to topple Fidel Castro's regime. Tensions between the U.S. and Cuba, sparked by Castro's rise to power in 1959, prompted CIA-backed plans to oust him. Brigade 2506, trained by the CIA, launched an invasion of Cuba but faced strong opposition from Cuban forces. The U.S. withdrew air support, resulting in the surrender of the invaders within three days. This debacle reinforced Castro's authority, deepened the gulf between the U.S. and Cuba, fueled anti-U.S. sentiments in Latin America, and heightened Cold War tensions, ultimately leading to the Cuban Missile Crisis in 1962.

Groupthink, a concept explored by Irving Janis in his book "Victims of Groupthink," illustrates the tendency for groups to conform, leading to flawed decision-making. This phenomenon was evident in the Bay of Pigs incident, which saw a rise in situations like mass suicides and biased jury verdicts. Groupthink fosters a false consensus and is considered detrimental, making healthy groups inefficient and irrational. Janis identifies cohesiveness, isolation, biased leadership, and decisional stress as the causes of this erroneous decision-making.

In the Bay of Pigs incident, a misguided Kennedy inherited a Cuban exile force trained by the CIA for an invasion and was now faced with the dilemma of addressing the increasingly visible exile force or pursuing specific foreign policy goals regarding Cuba. Kennedy's newness in office and inherited circumstances contributed to a lack of direction for the advisory group, setting the stage for flawed decision-making.

Groupthink occurred in the case of the Bay of Pigs due to the advisors' conformity, resulting in flawed decision-making, a false consensus, and the failure to challenge assumptions or consider crucial information.

Groupthink's Impact on Decision-Making

Groupthink, a widely studied psychological phenomenon, reflects how groups prioritize agreement over critical thinking, often resulting in flawed decision-making and the dismissal of differing perspectives. This tendency, outlined in *verywellmind.com* by Kendra Cherry, MSEd, a psychosocial rehabilitation specialist and psychology educator, suppresses individual doubts or reservations by emphasizing unanimity.

Cherry illustrates the significant impact of Groupthink on pivotal decisions, citing historical events like the escalation of the Vietnam War and political instances. This phenomenon, she emphasizes, leads to neglecting crucial information and stifling creativity within group dynamics. Distinguishing Groupthink from conformity, Cherry clarifies that while conformity aligns actions with a group, Groupthink specifically affects decision-making processes. However, she notes conformity can contribute to Groupthink without being the sole instigator.

Cherry outlines the pitfalls of Groupthink, which encompass disregarding dissenting opinions, lacking creativity, ignoring vital information, fostering overconfidence, and resisting new ideas. Offering preventive measures, she suggests fostering an environment that encourages critical thinking and dissent, assigning a "devil's advocate" role to challenge prevailing ideas, seeking external opinions, and promoting diversity among group members to encourage a broader range of perspectives.

"On the other hand, a good mood makes us more likely to accept our first impressions as true without challenging them."-**Daniel Kahneman.**

In our daily lives, we frequently face 'groupthink' effects, seen in decisions such as selecting a travel destination or determining the theme for a group performance. The majority occasionally make these choices without the group leader considering clear guidelines or budget constraints. When planning lacks a defined scope, execution might lack inclusivity or offer only partial benefits. Inclusive Leaders are crucial in resolving these situations by providing clarity and implementing comprehensive decision-making processes.

Influence of Cognitive Functions

The diversity of opinions among individuals can be attributed to their cherished values. As the adage goes, like-minded individuals tend to gravitate toward each other, forming alliances based on shared natural preferences, often leading to the dominance of the majority viewpoint. Studies reveal a varied distribution of personality types worldwide. Sensors constitute the majority, while intuitive feelers comprise only around 20%, with certain personality types particularly rare.

Consequently, our world's distribution of intelligence types—Practical, Analytical, and Creative—might differ accordingly. We inherently tend to endorse ideas proposed by individuals aligned with our inclinations or values. This alignment often leads us to support opinions or actions that resonate closely with our beliefs. This inclination towards the familiar can introduce a bias known as cognitive bias in the field of psychology.

Do you see how great ideas often lose out to less impressive ones backed by more people? Why does luxury often outweigh minimalism? Talented individuals sometimes get sidelined by those who excel at self-promotion. It's a constant battle between substance and surface. Let's dig into cognitive bias to explore this further.

"All of us are not always smarter than one of us; leaders need to distinguish between the wisdom of crowds and the madness of crowds. Leaders need to correct for cognitive biases the way a sharpshooter corrects for wind velocity or a yachtsman corrects for the tide."- **Paul Gibbons.**

Cognitive Bias

Cognitive biases represent consistent judgmental errors departing from rational thinking, stemming from individuals constructing their subjective realities. These biases lead to distorted perceptions, inaccurate conclusions, and sometimes irrational decisions. While some biases aid in faster decision-making and adaptability, others result from cognitive limitations. These biases, identified over decades, are vital in finance, management, and clinical judgment, offering insights to improve decision-making processes and strategies.

Cognitive biases significantly impact human thinking, decision-making, and beliefs. As detailed in the article from *verywellmind.com*, numerous biases shape our cognition. Here are some of the most prevalent ones:

1. Confirmation Bias: Favors information that confirms existing beliefs.
2. Hindsight Bias: Seeing events as more predictable after they've occurred.

3. Anchoring Bias: Excessively swayed by the initial information received.
4. Misinformation Effect: Memories influenced by post-event details.
5. Actor-Observer Bias: Attribute our actions to external influences and others' actions to internal causes.
6. False Consensus Effect: Overestimating others' agreement with our own beliefs.
7. Halo Effect: Initial impressions influence the overall perception of a person.
8. Self-Serving Bias: Taking credit for successes but blaming failures on external factors.
9. Availability Heuristic: Estimating probability based on easily recalled examples.
10. Optimism Bias: Overestimating the likelihood of good events and underestimating negative ones.

There are also other cognitive biases, such as the status quo bias (desire to maintain the current situation), apophenia (tendency to perceive patterns in random occurrences), and framing (presenting a position to give a particular impression). These biases collectively shape and influence human thinking patterns.

Understanding various cognitive biases helps in more conscious judgments. Recognizing biases in others, especially in leadership, aids in better decision-making. Leaders acknowledging these biases foster open-mindedness, inclusivity, and informed choices, mitigating biased outcomes and encouraging diverse perspectives.

In the article "Beware the Dangers of Cognitive Bias" by Pamela Rollings-Mazza, MD, BSN, CCHP, and Tommy Williams, BSN, RN, CCHP, unconscious biases shape our understanding through personal experiences. Cognitive biases like confirmation bias and gender bias, common among healthcare professionals, lead to erroneous diagnoses and flawed treatment decisions. Awareness and strategies, such as seeking diverse perspectives and questioning assumptions, combat biases, foster informed choices, and improve patient outcomes. Understanding and practicing bias awareness in healthcare promotes collaborative and effective decision-making processes.

Avoiding Cognitive Bias

In an article Authored by Timothy J. Legg, Ph.D., PsyD, and Rebecca Joy Stanborough, MFA, in *Healthline,* the bottom line emphasizes that cognitive biases can lead to inaccurate conclusions by disproportionately focusing on specific information. Although eradicating biases might be impractical, recognizing vulnerable situations and employing strategies like understanding biases, deliberate decision-making, collaboration, and objective processes can substantially diminish their impact.

In decision-making, it's challenging to entirely dodge cognitive biases due to the mind's inclination towards efficiency. However, researchers believe recognizing

situations prone to biases and implementing corrective measures is possible. The approach to mitigate these biases includes:
1. **Learning:** Studying cognitive biases aids in identification and counteraction once they're recognized.
2. **Questioning:** In situations vulnerable to bias, slow decision-making, and diversifies sources for more reliable information.
3. **Collaborating:** Engage diverse experts to explore perspectives beyond your own experiences.
4. **Remaining Blind:** Limit exposure to easily stereotyped factors like gender or race to minimize influence.
5. **Using Objective Measures:** Employ checklists, algorithms, and objective tools to focus on pertinent factors, reducing susceptibility to irrelevant ones.

Understanding oneself and dominant cognitive functions guides thinking patterns and biases. Feelers benefit from incorporating logic, while thinkers should develop emotional intelligence for empathy. Exploring sensor and intuitive perspectives enriches cognitive approaches. Individuation involves acknowledging shadows and integrating contradictions consciously.

"Knowing your own darkness is the best method for dealing with the darknesses of other people." - **Carl Jung.**

Critical Thinking

Critical thinking involves assessing available facts, evidence, observations, and arguments to form reasoned judgments using rational, skeptical, and impartial evaluation. This skill encompasses self-directed, disciplined, monitored, and corrective mental habits, serving as a means to appraise data. According to Richard W. Paul, critical thinking engages both intellectual abilities and personal traits. It demands strict standards of excellence, adept communication and problem-solving, and a commitment to surpass egocentric and sociocentric tendencies.

"Learn to use your brain power. Critical thinking is the key to creative problem-solving in business."- **Richard Branson.**

In his article on *Forbes.com*, contributor Bryce Hoffman emphasizes the importance of cultivating robust critical thinking skills in businesses today. This skill is essential for decision-making and problem-solving amid technological advancements and economic uncertainties. Employees armed with critical thinking skills can analyze information comprehensively, fostering better choices and innovative solutions. Their ability to assess facts and communicate effectively enhances collaboration and productivity. Lack of such skills leads to misunderstandings and poor decisions. Investing in training for these skills equips a workforce to navigate challenges adeptly, fostering efficiency, innovation, and productivity, driving organizational success.

Measurement of Critical Thinking skills

Measuring critical thinking skills involves considering both cognitive abilities and personality traits. Research links critical thinking to cognitive capacity, but the exact personality traits contributing to this skill are yet to be clearly defined.

In one study, 101 college students completed tests measuring critical thinking (using the Watson-Glaser Critical Thinking Appraisal), cognitive ability (specifically, three subtests from the Wechsler Adult Intelligence Scale-Third Edition), and personality traits (using the revised NEO Personality Inventory). Results showed that 'Openness to Experience' scores significantly affected critical thinking beyond cognitive abilities like understanding similarities.

Another study involving 105 students found similar results. Even after considering cognitive factors like verbal comprehension measured by the Verbal Comprehension Index, 'Openness to Experience' continued to influence critical thinking.

These findings suggest that fostering 'Openness to Experience' might enhance critical thinking skills among college students.

Improving Critical Thinking Skills

HBR notes that critical thinking is a learned behavior and provides three simple techniques to enhance this skill: questioning assumptions, applying logical reasoning, and broadening perspectives. While these practices may seem apparent, intentionally developing these three mental habits improves one's clear and sound reasoning ability.

Here are some characteristics and skills associated with critical thinkers:

1. **Analytical Skills**: The skill to deconstruct intricate matters into smaller elements to grasp their connections and consequences.
2. **Problem-Solving Skills**: Capacity to identify, define, and solve problems using critical thinking processes.
3. **Curiosity and Creativity**: A natural curiosity to explore, question, and generate innovative ideas or solutions.
4. **Information Literacy**: Proficiency in evaluating information sources' credibility, relevance, and reliability.
5. **Communication Skills**: Capability to articulate thoughts, ideas, and reasoning clearly and effectively in various formats (oral, written, visual).
6. **Logical Reasoning**: Proficiency in applying logic and reasoning to evaluate arguments, identify fallacies, and draw valid conclusions.

7. **Resilience and Persistence**: The ability to persist in problem-solving despite challenges or setbacks and willingness to revisit and refine strategies.
8. **Reflective Thinking**: Capacity for introspection and self-assessment to continuously improve one's critical thinking abilities.
9. **Empathy and Understanding**: Ability to empathize and understand diverse perspectives, fostering better collaboration and problem-solving in various teams or communities.
10. **Ethical Decision-Making**: Considers ethical implications and consequences while making decisions or analyzing problems.
11. **Risk Assessment**: The capacity to evaluate risks, consider outcomes, and make knowledgeable decisions grounded in calculated risk assessment.
12. **Meta-Cognition**: Being aware of one's thinking processes, monitoring cognitive biases, and adapting thinking strategies accordingly.
13. **Adaptability and Flexibility**: Adapting to new information, changing circumstances, and embracing different viewpoints.
14. **Strategic Thinking**: Capacity to think strategically, foreseeing long-term implications of decisions and actions.

15. **Cognitive Flexibility**: Ease in shifting thinking between multiple concepts or perspectives, allowing for a broader understanding of complex issues.
16. **Humility in Observation**: The humble approach to observation, emphasizing a willingness to learn from others while keenly observing and analyzing details within different contexts.
17. **Emotional Independence**: The ability to maintain emotional stability and make decisions based on rationality rather than being swayed solely by emotions or external pressures.
18. **Freedom from Social Validation:** The capability to make judgments and choices grounded in personal analysis and values, prioritizing individual convictions over the need for constant affirmation or conformity to societal norms or peer influence.

Mindset Practice

Recognize your cognitive biases by reflecting on particular situations, and assess how employing critical thinking might have altered the outcome.

4. Benefits of Thinking Slow

"**Logic, The art of thinking and reasoning in strict accordance with the limitations and incapacities of the human misunderstanding.**"**-Ambrose Bierce.**

The Toyota Production System (TPS) aims for waste elimination, emphasizing efficiency and "Just-in-Time" manufacturing. Rooted in "jidoka" (automation with human oversight) and JIT principles, TPS ensures rapid, high-quality vehicle production tailored to customer demands. TPS's focus on producing what's necessary when needed prevents defects and maximizes efficiency. This approach gives Toyota a competitive edge, serving as a cornerstone for cost reduction and continuous improvement. Toyota's commitment to refining TPS and nurturing its workforce underscores its pursuit of crafting exceptional cars that meet customer expectations, crucial for its enduring success and customer satisfaction.

Toyota's integration of JIT and Jidoka is synergistic. JIT aims to streamline production flow and reduce waste, while Jidoka ensures that quality is maintained at every step, empowering the workforce to intervene promptly in case of abnormalities. This logical combination contributes to Toyota's reputation for high-quality, efficient manufacturing and is a foundation for its continuous improvement initiatives.

Logic prevents inefficiency, damages, and wastage, as we see in the case of Toyota. But, from 2000 to 2010, Toyota vehicles experienced sudden unintended acceleration incidents linked to numerous deaths and injuries. Initially inconclusive, investigations later revealed in 2008 that a loose driver-side trim and problematic floor mats were causing acceleration issues. Toyota faced accusations of knowing about these problems but misleading consumers and continuing to manufacture flawed cars. In 2014, the Department of Justice imposed $1.2 billion in penalties on Toyota for mishandling safety concerns and deceptive practices regarding vehicle safety.

Logical thinking, while valuable and necessary, does indeed have its limitations. It's a systematic approach that helps problem-solving and decision-making by following a coherent and rational path. However, it may only sometimes guarantee a perfect or complete solution on the initial attempt. Factors like incomplete information, biases, or unforeseen variables can influence the outcome, leading to potential errors or oversights. While logical thinking is a robust tool, it's flexible and might require revisions or iterations to reach the most accurate or effective conclusions, especially in complex or uncertain situations. "Pure logical thinking cannot yield us any knowledge of the empirical world. All knowledge of reality starts from experience and ends in it."- **Albert Einstein.**

Have you ever taken competitive exams? When repeatedly tackling those questions, do you often get most of the answers wrong? As you invest more time and become familiar with how the questions intend you to think and respond, you can approach problems more effectively. What was the turning point? You acquired the technique to navigate each situation where the questions attempted to deceive you. Does this imply that we can become more logical through training and experience?

Consider the case of Toyota. After diligently analyzing where they made mistakes, would they be better prepared for unforeseen circumstances?

Benefits of Thinking Slow

Eva M. Krockow, Ph.D.'s Psychology Today article, "The Benefits of Being a Slow Thinker," explores the concept of deliberate, unhurried thinking for improved decision-making. Taking time to process information can result in fewer errors and reduced bias. Rapid thinking often relies on intuition, whereas deliberate thinking involves a careful, meticulous approach to decision-making. This method helps mitigate biases and leads to better outcomes.

The article discusses the Cognitive Reflection Test, revealing that initial instinctual responses may not always be the most accurate. Quick, intuitive answers sometimes require correction, particularly when faced with tricky questions. It doesn't discredit fast thinking but highlights how measured, thoughtful deliberation can prevent errors from impulsive judgments. It emphasizes that neither slow nor fast thinking is universally superior. Instead, it underscores the importance of recognizing when each approach is beneficial. Simple choices, like daily attire, may warrant rapid thinking. However, significant decisions, such as selecting an elective, benefit from careful consideration. The critical takeaway is to avoid favouring one thinking style over the other but to discern when to employ each method. Balancing deliberate, intuitive, and rapid thinking contributes to better decision-making, leveraging the strengths of each approach as per the demands of a given situation.

"When intuition and logic agree, you are always right." - **Blaise Pascal.**

The Dilemma of Logical Decision

"Directed by Ben Affleck, 'Gone Baby Gone' is a gripping neo-noir crime thriller based on Dennis Lehane's novel. When four-year-old Amanda McCready goes missing in Dorchester, Boston, private investigators Patrick Kenzie and Angie Gennaro are hired by the girl's troubled and drug-addicted mother, Helene McCready, to find her. Revelations about a staged kidnapping, the ulterior motives of the family members and the police, and personal guilt shake Kenzie's moral compass to the core. Amidst this turmoil, Kenzie faces a monumental decision upon discovering that Amanda was happily taken care of by another family—to uphold the promise he made to Helene to find Amanda or look the other way, thus enabling a better future for Amanda. Unable to break his promise to Helene, Patrick notifies the police about Amanda's whereabouts, believing she belongs with her mother, regardless of Helene's parenting. Police chief Doyle and Amanda's uncle Lionel, irrespective of their good intentions as they conflicted with the law of the land, are arrested, while Patrick and Angie break up.

In the climax, Patrick visits Helene as she prepares for a date. He learns she hasn't planned appropriately for a babysitter, so he volunteers to watch Amanda. After Helene leaves, Patrick asks Amanda about her doll, Mirabelle, but she insists the doll's name is Anabelle. Patrick and Amanda then sit in silence, watching TV.

The film's climax is a poignant reflection on morality, justice, and the complexities of human nature. As the credits roll, audiences are left pondering the haunting choices made by the characters and the relentless grip of moral ambiguity. Logically, the case is successfully closed, but whether it has served justice is beyond logic.

"A mind all logic is like a knife all blade. It makes the hand bleed that uses it." – **Rabindranath Tagore.**

Measuring how we think

The road to improvement begins with self-awareness, for as the saying goes, 'You can't improve what you don't measure,' often attributed to Peter Drucker.

The cognitive reflection test (CRT) is a tool to gauge an individual's inclination to challenge initial, intuitive responses and engage in deeper reflection to arrive at a correct answer. Introduced by psychologist Shane Frederick in 2005, the CRT's validity as a measure of "cognitive reflection" or "intuitive thinking" has faced scrutiny. Although it moderately correlates with intelligence tests like the Intelligence Quotient (IQ) test and aligns significantly with various mental heuristic measures, debates persist regarding its actual assessment of cognitive abilities or intelligence.

Further research has revealed that CRT encompasses multiple facets: while some individuals immediately arrive at the correct answer, others struggle even after reflecting on their initial intuition. Moreover, experts argue that merely suppressing the first response does not solely contribute to successful CRT performance; factors such as numeracy and reflectivity also influence results.

At its core, CRT operates based on two cognitive systems coined "system 1" and "system 2," a concept initially introduced by Keith Stanovich and Richard West. System 1 functions swiftly without conscious deliberation, whereas System 2 demands deliberate, conscious thinking. This test comprises three questions, each presenting an obvious yet incorrect response from System 1. However, arriving at the correct answer necessitates the activation of system 2. This activation relies on recognizing the initial error, prompting individuals to reflect upon their thought process. The CRT's connections to other measures provoke discussions regarding its assessment and its link to cognitive abilities and intelligence.

"Measurement is fabulous. Unless you're busy measuring what's easy to measure as opposed to what's important."- **Seth Godin.**

Logical Reasoning

Logical reasoning is a methodical mental process aiming to conclude from premises through rigorous arguments. It revolves around propositions—statements that can be true or false—and follows specific norms to construct convincing reasoning. This process is studied under the discipline of logic.

Different types of logical reasoning vary in their certainty of conclusions. Deductive reasoning offers the most robust support, ensuring that if the premises are true, the conclusion must also be proper. An example is, "All men are mortal; Socrates is a man; therefore, Socrates is mortal." Deductive reasoning is foundational in formal logic and math.

Non-deductive reasoning doesn't guarantee the truth of conclusions but makes them rationally convincing.

Inductive reasoning infers general laws from observed patterns, while abductive reasoning seeks the best explanation for a given observation. Analogical reasoning compares similar systems to make inferences.

Logical fallacies represent flawed reasoning. They can be formal, based on the structure of an argument, or informal, rooted in the content or context. In a broader sense, logical reasoning involves critical thinking skills such as assessing information reliability, generating and evaluating reasons, seeking new information, avoiding inconsistencies, and considering different options before making decisions.

Logical reasoning is crucial in academic study and everyday decision-making, bridging the gap between premises and conclusions through systematic and convincing arguments.

"When dealing with people, remember you are not dealing with creatures of logic, but creatures of emotion." – **Dale Carnegie.**

Reasons why logic is crucial in daily life

Logic, an indispensable framework guiding rational decision-making based on evidence, significantly elevates problem-solving, interpersonal relationships, and overall well-being. In every facet of life, logical thinking is the compass steering decisions toward truth and evidence over emotional sway. Mastering logic in professional spheres such as law, accounting, business, and HR provides a distinct advantage by enhancing persuasiveness, communication skills, and problem-solving prowess. Relying on logic fosters confidence in decision-making by anchoring choices in concrete facts. It fortifies persuasive abilities and expedites problem-solving by focusing on factual foundations, minimizing distractions, and expediting resolutions.

In citizenship matters, logic fosters ethical decision-making by discerning between right and wrong, nurturing a framework for moral actions. Understanding fallacies is crucial for honing critical thinking skills, and the absence of logic significantly hampers problem-solving, impacting one's quality of life.

Logic contributes to stress management by providing a broader perspective that reduces stress by dissecting events logically. By mitigating emotional biases, logic enables more reasoned conclusions, diminishing the influence of emotions in response.

In conflict resolution, logic is a bulwark against emotional arguments, preserving peace in discordant situations. Its role in enhancing interpersonal skills through effective communication and removing emotional sensitivity from decision-making processes further emphasizes its significance in daily life.

Key Behaviours of a Logical Thinker

An article in *powerofpositivity.com* discusses twenty key behaviours that characterize a logical thinker:

1. Paying close attention to details, even in familiar or unfamiliar situations.
2. Willingness to admit when they're incorrect or lack knowledge.
3. Ability to convey ideas effectively through appropriate language.
4. Openness to new ideas and differing opinions for a comprehensive view.
5. Prioritizing having all relevant facts and testing ideas for accuracy.
6. Avoiding hasty conclusions by thoroughly gathering information.
7. Steering clear of evasive language to ensure transparency.
8. Engaging in critical thinking and self-examination, asking necessary questions.
9. Presenting data clearly after testing ideas for accuracy.
10. Refusing to accept information at face value, seeking comprehensive understanding.
11. Proficiency in effective communication, explaining thoroughly based on facts.
12. Maintaining distinctive habits and adhering to plans consistently.
13. Tracking the origins and history of ideas for sound decision-making.

14. Avoiding over-analysis by balancing information gathering and decision-making.
15. Striving for truth based on solid evidence or proof.
16. Procrastinating due to a desire for thorough information before conclusions.
17. Debating ideas to improve understanding rather than arguing.
18. Continuously learning and staying well-informed on various topics.
19. Avoiding vague explanations, preferring precision in communication.
20. Always have a plan driven by questioning and pattern recognition.

The article emphasizes that developing these behaviors takes time and practice but is essential for making better decisions and building beneficial habits. Logical thinking enhances problem-solving, decision-making, and productivity.

Mindset Practice

Assess your logical thinking behavior and areas you need to improve.

5. Successful Problem Solving

"Successful problem-solving requires finding the right solution to the right problem. We fail more often because we solve the wrong problem than because we get the wrong solution to the right problem."- Russell L. Ackoff.

In 2011, Sony faced a daunting challenge: its consumer electronics arm, particularly the TV business, was hemorrhaging money. Kazuo Hirai inherited this turbulent landscape with a critical realization—the company's strategy, fixated on boosting TV sales to offset high operational costs, was fundamentally flawed. An in-depth analysis confirmed a stark truth: Sony needed to sell a staggering 40 million TV sets annually for viability, yet in 2010, sales languished at just 15 million units.

Hirai diverged from conventional wisdom, steering Sony away from a volume-centric approach. Instead, he directed efforts towards raising prices and quality, slashing LCD TV sales in developed markets by 40%, and streamlining the U.S. product lineup significantly. They initiated cost restructuring measures and focused on enhancing picture quality.

This strategic pivot bore fruit in 2015 as Sony's TV division celebrated its first operating profit in 11 years. Hirai's decisive leadership, marked by a shift in sales strategy, cost efficiencies, product enhancement, and innovative retail tactics, redefined Sony's trajectory and stood as a paragon of corporate revitalization in the business world.

Sony's problem was their product's inability to generate profits. Initially, they focused on boosting sales volume as the solution. However, the issue was resolved by addressing increased cost per unit, improving product quality, and implementing innovative retail solutions. These strategies effectively tackled Sony's profitability problem.

Are you solving the right problems?

In the quest for practical problem-solving, reframing plays a pivotal role. A case study discussed in the HBR guide explains a situation to better identify solutions for a problem. It is in a scenario called the slow elevator problem.

Picture owning an office building with disgruntled tenants complaining about a slow elevator. Conventional solutions—replacing the elevator or enhancing its mechanisms—occupy the typical solution space, assuming speed as the core problem. However, building managers introduced an unconventional yet ingenious idea: installing mirrors near the elevator. This solution doesn't directly address the speed but captivates people's interest, reducing complaints by altering their perception of time. This exemplifies reframing—a shift from conventional problem definitions to uncovering innovative solutions. It involves questioning assumptions, seeking diverse perspectives, and exploring unconventional ideas. Reframing encourages the discovery of underlying issues like peak demand or human behavior, enabling the identification of more impactful solutions. Continual reassessment and experimentation are critical, as reframing isn't a one-time fix but a dynamic process fostering innovative problem-solving beyond initial perceptions.

The slow speed of the elevator was a concern, but the real issue for its users was the extended waiting time. They found an equivalent solution to the problem in the mirrors, thereby addressing the wait duration.

Pitfalls of Problem-solving in Business

The article in Forbes titled "The Five Pitfalls of Problem-Solving - And How to Avoid Them" highlights critical errors encountered in complex problem-solving scenarios:

1. **Flawed Problem Definition**: Businesses need to define problems consistently. It stresses the importance of clearly defining problems using the TOSCA checklist, which assists in a comprehensive understanding of practical solutions.

2. **Solution Confirmation**: Rushing into solutions requires understanding the problem thoroughly to avoid failed ventures. Breaking down core questions into sub-questions is advised for a deeper investigation of potential solutions.

3. **Wrong Framework**: Applying frameworks with consideration of crucial aspects of a problem can be beneficial. Caution is advised when using frameworks, ensuring they align with the specific issue.
4. **Narrow Framing**: Superficially understanding a complex problem and drawing quick analogies can lead to misguided solutions. It advocates for a design thinking approach that comprehensively understands issues from the users' perspective.
5. **Miscommunication**: Ineffective communication of solutions can hinder progress, even if based on accurate information. It emphasizes presenting well-grounded, evidence-backed solutions from a structured problem-solving process.

The article concludes by proposing the 4S method as a disciplined problem-solving approach, guiding individuals through problem-stating, structuring, solving, and selling. This method underscores the significance of structured problem-solving processes in achieving successful outcomes, as outlined in the book, "Cracked It! How to Solve Big Problems and Sell Solutions Like Top Strategy Consultants." Successful business problem-solving is towards achieving profits or meeting specific business objectives.

However, in our daily lives, success in problem-solving is relative. It requires an integrated approach- practical, analytical, and creative.

"We cannot solve our problems with the same level of thinking that created them."-**Albert Einstein**.

Typical Problems According to Personality Types

We encounter individuals with diverse personality types characterized by unique cognitive functions and values. Their predominant intelligence varies, leading them to meet specific challenges based on the tasks in which they lack expertise. For instance, introverts may grapple with different issues compared to extroverts. Recognizing these challenges and devising solutions can significantly aid us.

As Oprah Winfrey aptly said, "Everyone has to learn to think differently, bigger, to be open to possibilities."

Every individual naturally tends to operate according to their primary cognitive functions. However, personal growth becomes manageable if one breaks free from the confines of their default type. Remaining within the comfort zone can make it challenging to relate to others and impede personal development.

World Economic Forum (WEF) often explores the impact of behavioral sciences on economies, industries, and global issues. One popular yet controversial tool in the business world for understanding personalities is the Myers-Briggs Type Indicator (MBTI). This assessment categorizes individuals into 16 personality types based on four behavioral binaries.

Despite its controversy, the MBTI holds significant sway, with approximately 80% of Fortune 500 companies and 89 of the Fortune 100 companies utilizing it to evaluate their employees. The Career Assessment Site has developed an informative infographic highlighting each personality type's strengths and weaknesses and average education level, along with details on the prevalence of each type and their average household income. Understanding how our kind is faring is a tool that can be used to plan for the future and probable challenges that can arise concerning the weaknesses of the personality type.

Applying MBTI Advantage for Success

Understanding your MBTI results can offer many opportunities for self-discovery and improvement, mainly when applied in various aspects of your life. Forbes Coaches Council members provide valuable advice on harnessing these insights at home and in the workplace.

Here are some key ways to apply and benefit from your MBTI results:

1. **Studying Behavioural Impact:** Use your MBTI results to become more self-aware of how your behavior influences others. Tailor your communication and interactions accordingly.
2. **Structuring Your Day:** Recognize your introvert/extrovert tendencies to organize your day

more effectively, acknowledging your need for social interaction or solitude.
3. **Appreciating Differences:** Embrace diversity by understanding and appreciating the perspectives of individuals with different personality types.
4. **Creating a Development Plan:** Identify strengths and weaknesses revealed by your MBTI results to chart a personal or professional development path.
5. **Addressing Defensive Reactions:** When your personality assessment triggers defensive reactions, it signals areas for personal growth and improvement.
6. **Enhancing Communication Strategies:** Utilize MBTI insights to refine your communication strategies, catering to others' needs and communication styles.
7. **Managing Preferences:** Acknowledge your preferences without feeling limited by them, driving them effectively for better outcomes.
8. **Building Relationships:** Foster stronger connections by understanding how your type influences your interactions and openly sharing this information with others.
9. **Navigating Career Paths:** Utilize MBTI to explore potential career paths based on identified strengths and preferences.

10. **Setting Clear Expectations:** Communicate your working style to teammates to manage expectations and foster a harmonious work environment.
11. **Proactive Communication:** Share your MBTI results openly to guide others on how best to collaborate and communicate with you.
12. **Improving Leadership Style:** Use MBTI insights to refine leadership behaviors, enhancing your impact and support for your team.
13. **Mitigating Conflict:** Increased awareness of personality differences can minimize conflicts by fostering empathy and understanding among team members.

Leveraging your MBTI results isn't just about self-discovery; it's a personal and professional growth tool. By embracing these insights, you can navigate relationships, communication, and career paths more effectively, fostering a more harmonious and productive environment at work and in your personal life.

An analytical mindset involves considering statistics, trends, and facts for future planning. Anticipating and addressing potential issues is a successful problem-solving approach.

"An idiot with a plan can beat a genius without a plan."- **Warren Buffet.**

Problem-Solving Strategies

Mastering multiple strategies is crucial as distinct problems demand different approaches, empowering individuals to select the most fitting plan when faced with challenges. This versatility expedites problem-solving and fortifies critical thinking skills.

Diving into various problem-solving strategies, let's explore some examples listed on indeed.com:

1. **Defining the Problem:** Carefully breaking down challenges aids in recognizing their extent and devising targeted solutions. For instance, a company struggling with high employee turnover may discover the issue lies in ineffective onboarding rather than solely in hiring.

2. **Visualizing and Diagramming the Problem:** Visualizing complex problems or drawing diagrams enhances understanding. Whether fixing a printer or streamlining a product development process, visual aids clarify the issue's elements.

3. **Breaking Down Complexity:** Dividing more significant problems into smaller, manageable tasks streamlines resolution, allowing focused attention on individual elements.

4. **Redefining the Problem:** Sometimes, redefining a problem uncovers alternative perspectives and solutions. For instance, a business seeking to meet

high consumer demand may opt for a different product that aligns with available resources.

5. **Collecting and Organizing Information:** Organizing problem-related data into charts or tables helps discern patterns, as seen when predicting the lifespan of a product like a laptop.

6. **Working Backward:** Reverse engineering or retracing steps aid in reproducing successful solutions, as demonstrated in creating a superior product by analyzing a competitor's model.

7. **Using Problem-Defining Methods:** Employing techniques like the Kipling Method by asking specific questions helps define problems and structure approaches.

8. **Leveraging Past Experience:** Drawing from past experiences aids in adapting successful approaches to current problems, which is beneficial for marketing campaigns or similar endeavors.

9. **Engaging Facilitators:** Introducing neutral facilitators during complex problem-solving sessions enhances productivity and fosters effective collaboration.

10. **Trial-and-Error Approach:** Experimenting with various solutions and assessing their effectiveness assists in identifying the most suitable approach.

11. **Decision Matrix Evaluation:** Ranking potential solutions based on predetermined criteria aids in selecting the most viable option through weighted evaluations.
12. **Seeking Peer Input:** Soliciting opinions from diverse peers exposes individuals to fresh perspectives and unique solutions, enhancing problem-solving.
13. **Taking Breaks:** Temporarily stepping away from a problem provides mental rejuvenation, fostering clearer thinking and renewed focus upon return.

Mastering these problem-solving strategies equips individuals with versatile tools to address challenges effectively across diverse scenarios. Brian Tracy delineates the below seven problem-solving stages in his book "Creativity and Problem Solving."

1. **Define Your Problem:** Write your problem clearly to engage your brain fully. Often, half the battle is won by precisely defining the issue.
2. **Read, Research, and Gather Information**: Gather as many facts as possible about the problem. Avoid falling in love with an initial solution and validate assumptions by seeking information from diverse sources.
3. **Don't Reinvent the Wheel**: Recognize that many problems have already been solved. Consult

experts, ask informed individuals, and learn from others who have faced similar challenges.

4. **Let Your Subconscious Work**: After assembling and discussing information, consciously attempt to solve the problem. If you're unsatisfied with the solutions, set it aside and revisit it later. Sometimes, letting your subconscious mind work on it can provide unexpected solutions.

5. **Use Your Sleep**: Review the problem before sleeping and ask your subconscious mind for a solution. Often, your subconscious works on problems during sleep, offering insights or solutions upon waking.

6. **Write It Down**: Keep a notepad handy to jot down insights or solutions that come to you anytime. Remember valuable ideas that could save time and effort.

7. **Take Action**: Act promptly on the ideas or solutions generated. Ideas can be time-sensitive so that immediate action can lead to positive outcomes.

Mindset Practice

Reframe a problem, apply a problem-solving strategy, and watch the outcome.

6. Strategies for Right Decisions

"The best decision-makers are always armed with the best information and data!"- George Raveling.

Netflix recently divulged astonishing statistics, revealing that viewers consumed nearly 92 billion hours of content during the first half of 2023, equivalent to roughly 11,000 years of TV. The disclosed data showcased specific viewing numbers for the platform's most popular shows, with Netflix's own production, "The Night Agent," claiming the top spot at 812 million viewing hours, followed by "Ginny & Georgia" Season 2 at 665 million hours. Surprisingly, 18,129 shows amassed over 50,000 viewing hours each.

The analysis highlights that people predominantly engage with recent content, with roughly one-third of viewing hours spent on shows released in 2022 and 2023. Most top-viewed programs were also from these recent years, with English-language content leading, followed by Korean and Japanese productions. Notably, around 30% of viewing occurs in non-English languages.

Additionally, Netflix revealed insights into prevalent words in show titles, indicating genres such as romance, drama, and action as primary themes. The data set, comprising over 18,000 titles, covers 99% of Netflix viewing, with 60% of Netflix-produced shows making their weekly top 10 lists.

Moving forward, Netflix intends to release similar viewing data twice yearly, providing greater transparency into viewership trends. Utilizing data analytics, Netflix comprehends its users more deeply, customizes its content selection, and consistently enhances the user journey. Personalized recommendations stand out as a crucial element fuelling Netflix's triumph, empowered by the capabilities of big data.

Leveraging Big Data and data analytics is crucial in navigating economic challenges for business success. These tools offer insights from large data volumes, empowering agile decision-making and optimizing business growth. By identifying patterns, organizations fine-tune sales strategies, streamline processes, boost productivity, and refine customer profiling. Such analytics facilitate rapid adaptations to market changes, ensuring competitive pricing models and enhanced customer experiences. Despite implementation challenges, the rewards are substantial, with Gartner reporting 42% higher-than-expected sales analytics ROI. Emphasizing the importance of high data quality and nurturing a culture centered around data is crucial to leveraging the unquestionable advantages of Big Data and analytics in contemporary business models.

"Information is the oil of the 21st century, and analytics is the combustion engine."- **Peter Sondergaard.**

Data-Driven Decision Making

Data-driven decision-making (DDDM) revolutionizes business paradigms by anchoring choices on meticulous data analysis. It champions confidence through insights, empowering proactive strategies and cost efficiencies. Leveraging data reveals success stories, from Google's people analytics to Starbucks' location precision and Amazon's personalized recommendations. Embracing this approach fosters pattern recognition, anchors decisions to data, utilizes visual interpretation and advocates continuous learning. While intuition holds significance, integrating data enhances organizational potential incrementally.

While data has historically underpinned corporate strategies, the contemporary era churns out an astronomical 2.5 quintillion bytes of data daily. This abundance has democratized data accessibility for businesses of all scales, rendering data-driven decision-making a modern, prevalent phenomenon.

Benefits of Data-Driven Decision-Making

Data-driven decision-making epitomizes the process of anchoring decisions on meticulous data analysis and validation before their implementation. Businesses across industries leverage data in diverse ways – from mining customer preferences through surveys to deploying user tests for product evaluation and market analysis. Here are a few benefits of DDDM.

1. **Augmented Decision Confidence**: Data analysis engenders a robust foundation for confident decision-making. It benchmarks existing scenarios, fostering a deeper understanding of potential decision impacts and instills unwavering confidence in organizational strategies.
2. **Proactive Versatility**: Transitioning from reactive to proactive decision-making, empowered by data, enables businesses to identify opportunities ahead of competitors and pre-empt potential threats.
3. **Cost-Efficiency**: Harnessing data to curtail expenses is one of the most impactful aspects of data-driven initiatives. Many organizations have reaped value by leveraging data to enhance operational efficiency and make informed, cost-effective decisions.

Strategies for Embracing a Data-Centric Approach

Embracing data-driven decision-making doesn't necessitate an immediate, all-encompassing transformation. Starting small, benchmarking progress, meticulous documentation, and iterative adjustments pave the way for an incremental yet profound shift toward a data-driven ethos. Harvard Business School's Online Business Insights Blog provides the insights below.

1. **Pattern Recognition**: Developing an analytical mindset involves identifying patterns in diverse

datasets, fostering insights, and drawing meaningful conclusions. This practice transcends professional domains, permeating everyday life.

2. **Data-Driven Decision Loop**: Anchoring decisions on available data and cultivating a habit of data-driven decision-making reinforces the analytical approach, facilitating confident and informed choices.
3. **Visual Data Interpretation**: Embracing data visualization techniques aids in deciphering complex datasets, empowering individuals to glean insights and make informed decisions effortlessly.
4. **Continuous Learning**: Pursuing educational avenues such as data analytics courses or degrees augments one's data science proficiency, enabling a deeper data integration in decision-making processes.

"There were 5 exabytes of information created between the dawn of civilization through 2003, but that much information is now created every two days."- **Eric Schmidt.**

Noah Mitsuhashi, a former member of the Forbes Technology Council, discusses the critical role of data in making informed decisions in an article titled "Using Data To Make Better Decisions." He highlights the challenge of making accurate guesses without relevant background information, likening it to watching a sports game with only partial information. In this light, investors need to access all available data to make decisions. Mitsuhashi emphasizes the importance of data in simplifying complex ideas and making sense of vast numbers, advocating for data explanation and visualization as essential tools in understanding intricate data points. He points out that financial data, while complex, can be presented more comprehensibly through graphs and comparisons.

However, he stresses that not all data sources are reliable, citing examples of inaccurate and manipulated data in crypto, financial markets, and trading practices. Mitsuhashi warns against unquestioningly trusting data sources, advocating for careful evaluation and skepticism, mainly when critical decisions are based on them.

AI-based decision-making tools

An article by Haillie Parker introduces ten AI tools tailored for decision-making processes. The article emphasizes evaluating AI tools based on their data analysis capacity, accuracy, error detection, data privacy, generative capabilities, and flexibility before selecting the most suitable tool for specific business needs. Each tool caters to different aspects of decision-making, from data analysis to visualization, each possessing varying strengths and limitations.

1. **ClickUp**: Offers AI assistance for project management, brainstorming, generative writing, and decision-making frameworks.
2. **Athenic AI (Formerly AskEdith)**: Focuses on self-service data analytics for product and market data analysis.
3. **Baseboard**: Provides customizable data visualization without requiring coding experience.
4. **AI Consulting Tools**: Offers SWOT, PESTEL, lean canvas, and user persona reports to support HR and marketing decisions.
5. **Findly AI**: It uses a chatbot-like interface for data visualization, which is handy for e-commerce and healthcare data.
6. **PlotGPT**: Utilizes ChatGPT to analyze and customize data, offering insights and infographics across industries.
7. **Hal9**: Enables easy departmental analytics and visual edits, supporting data-driven decisions.
8. **Intellibase**: Focuses on understanding customer demands and trends for product prioritization.
9. **ChartAI**: Utilizes GPT-4 for diagram creation to facilitate process-based decisions.
10. **Smarter Sales**: Provides AI-generated feedback for customer support calls to improve decision-making regarding support initiatives.

"Without big data analytics, companies are blind and deaf, wandering out onto the web like deer on a freeway."- **Geoffrey Moore.**

Seven steps for effective decision-making.

The HBR guide to making better decisions gives the below seven steps for effective decision-making. It highlights the need to actively use this checklist to counter biases and improve decision outcomes. Research shows that managers who consistently apply these steps save time, make quicker decisions, and enhance outcomes by 20%, emphasizing the importance of a new, scalable approach to decision-making that acknowledges psychological biases and utilizes simple, impactful tools.

1. **Align with Company Goals:** Write down five preexisting company goals or priorities impacted by the decision to focus on what's crucial and avoid post-decision rationalization.
2. **Explore Alternatives:** List at least three, preferably four or more, realistic alternatives to expand choices and improve decision quality.
3. **Identify Missing Information:** Write down essential information you lack to prevent overlooking unknowns amidst known details prevalent in information-rich environments.
4. **Envision Future Impact:** Describe the decision's impact one year ahead to create a narrative, gaining perspective from similar scenarios.

5. **Involve Stakeholders:** Engage a team of two to six stakeholders to minimize bias and enhance decision buy-in, avoiding larger groups for optimal effectiveness.
6. **Document Decision and Support:** Record the decision, the reasons behind it, and the team's level of support to boost commitment and measure outcomes.
7. **Schedule Follow-Up:** Plan a follow-up within one to two months to track progress, make corrections, and learn from the decision's results, ensuring continuous improvement.

When consistently applied, these steps have shown significant time savings, quicker decision-making, and improved outcomes, highlighting their vital role in decision-making processes. As demonstrated in this chapter, leveraging data intelligence contributes to improving the outcomes of decision-making.

"Every company has big data in its future and every company will eventually be in the data business."- **Thomas H. Davenport.**

Frameworks for Intelligent Decision-Making

In her insightful article in Psychology Today, Allison E McWilliams, Ph.D., underscores the importance of finding personalized decision-making frameworks to counteract analysis-paralysis and disaster-fantasizing. She emphasizes that while perfect information is unattainable, understanding personal motivations, gathering sufficient information, and seeking guidance facilitate intelligent decision-making tailored to individual needs.

She introduces two powerful frameworks that aid in effective decision-making, emphasizing the importance of aligning decisions with personal motivations. The first framework, derived from manufacturing project management, is known as the "project management triangle." This framework prompts individuals to evaluate decisions based on three crucial factors: time, cost, and quality, where choosing two necessitates sacrificing the third. McWilliams illustrates this by emphasizing her preference for time, valuing it as her most precious commodity. She emphasizes her readiness to invest in services that save time, prioritizing efficiency over spending hours researching or bargain shopping.

Although originally tailored for project management contexts, the project management triangle is a valuable guide for day-to-day decisions. McWilliams encourages a focused approach, aligning choices with personal priorities by urging individuals to assess whether a decision revolves around time, cost, or quality.

The second framework McWilliams presents is rooted in "intelligent careers" research, emphasizing three core competencies: knowing why, knowing how, and knowing whom. This framework addresses more considerable career and life decisions, stressing the importance of understanding personal values and skills required for current and future roles and identifying a supportive network for guidance.

Highlighting the significance of self-awareness, McWilliams encourages individuals facing career or life decisions to ask themselves critical questions about their motivations (knowing why), knowledge (knowing how), and available support systems (knowing whom). By employing these frameworks, individuals can navigate decisions effectively, considering personal motivations while seeking relevant information and guidance.

McWilliams acknowledges that no framework is foolproof, and decision-making can be complex. However, by comprehending personal motivations, gathering pertinent information, and seeking advice from mentors or counselors, individuals can make informed and intelligent decisions tailored to their unique circumstances.

"If you are making smart decisions and doing the right things in the correct order and dedicating yourself to something, the success is inevitable." - **Matt Snell.**

Addressing Wrong Decisions

In her article, "What to Do When You've Made a Wrong Decision," featured in HBR's guide to making better decisions, Dorie Clark addresses the challenging reality of acknowledging and rectifying bad choices. Admitting a misstep can be daunting, be it hiring the wrong person, choosing an ill-suited job, or launching an unsuccessful product line. Clark offers a strategic roadmap to navigate this scenario. She emphasizes the crucial need for swift action, cautioning against succumbing to the sunk cost fallacy. Despite invested time, money, or effort, it's essential to recognize when to cut losses rather than persisting with doomed endeavors. Identifying the remedy becomes pivotal. Some situations may warrant remedial measures, such as additional training for an underperforming hire or scaling back an unsuccessful expansion to gain insights. However, Clark notes that some blunders necessitate decisive action. Sudden resignation from an unsatisfactory job could pave the way for a better-suited candidate's opportunity. Extracting lessons from the misstep is imperative. Clark urges introspection to identify foreseeable issues, whether due to oversight, reliance on unreliable sources, or over-optimism, enabling individuals to learn and evolve from the experience.

Sharing knowledge about wrong decisions is powerful, despite the inclination to conceal them. Clark cites the example of Jared Kleinert, who openly discussed his failure in a Forbes article, earning respect for his transparency. Acknowledging mistakes, working towards resolutions, and openly sharing lessons learned mitigate initial errors and foster lasting respect among peers.

Clark's insights from the article encapsulate the importance of promptly acknowledging misjudgements, seeking remedies, extracting valuable lessons, and openly sharing experiences to navigate wrong decisions effectively, fostering personal growth and professional respect.

Mindset Practice

Employ a recommended decision-making framework and observe the outcome.

Conclusion

"Grow Analytical Mindset" emphasizes the pivotal role of critical thinking and logical reasoning in cultivating successful intelligence. It explores the trio of practical, analytical, and creative intelligence, acknowledging the diverse cognitive abilities exhibited across various personality types. The synthesis of these elements culminates in integrated thinking and collective intelligence, a notion echoed by Einstein, highlighting the inclusiveness of intelligence.

Backed by insights from neuroscientists and psychologists, the imperative nature of amalgamating cognitive abilities stands paramount. The book advocates for Idealists to refine their analytical intelligence while recognizing and leveraging their inherent strengths in creative intelligence. Specifically, it urges Idealists to bolster their analytical capabilities by embracing fact-finding and logical reasoning as areas for improvement.

Building on this groundwork, the subsequent installment in this series will explore cultivating a Creative Mindset. This upcoming book aims to guide readers in honing their creative potential, irrespective of their cognitive strengths. It seeks to foster a more holistic and balanced approach to problem-solving and decision-making.

GRAB YOUR FREE GIFT BOOK

MBTI enumerates 16 types of people in the world. Each of us is endowed with different talents, which prove to be the innate strength of our personality. To understand the deeper psychology of your personality type, unique cognitive functions, and integrated personality growth path, visit www.clearcareer.in for a free download –

"Your Personality Strength Report"

About the Author

Devi Sunny, a passionate author and mentor, has been fortunate to create the series: 'Clear Career Inclusive,' 'Fearless Empathy,' and 'Successful Intelligence.' She nurtures inclusive spaces, fosters empathetic leadership, and encourages cognitive growth. At Clear Career, she strives to offer guidance based on her experiences. If seeking supportive career insights, please reach out at contact@clearcareer.in.

May I ask for a Review

Thank you for taking out time to read this book. Reviews are the essential for any author. I look forward to your feedback and reviews for this book. I welcome your inputs to incorporate in and deliver an even better book in my next attempt in the very near future. Please write to me at:

contact@clearcareer.in

Your support will help me to reach out to more people. Thanks for supporting my work. I'd love to see your review and feel free to contact me for any clarifications.

Preview of Previous Books

Successful Intelligence Series

Book 1: Grow Practical Mindset

Are you prepared to elevate your adaptability, enhance practical problem-solving skills, and refine your judgment with practical intelligence?

Unlock the keys to practical intelligence with 'Grow Practical Mindset,' the inaugural book in the 'Successful Intelligence' series. Delve into the foundational principles of Sternberg's Triarchic Theory, focusing on adaptability and problem-solving skills crucial for success across diverse scenarios.
From practical exercises to insightful strategies, this book equips you with actionable tools to enhance decision-making in your daily life, empowering you to thrive in various environments.
This book will take you on an exhilarating journey through the following key topics, unraveling intriguing insights-

1. Practical Mindset
2. Triarchic Theory of Intelligence
3. Practical Intelligence
4. Examples of Practical Intelligence
5. 12 Traits for a Practical Mindset
6. Personality Types Natural Preferences
7. Cognitive Functions
8. Personality Growth
9. Fixed Mindset
10. How to Overcome a Fixed Mindset
11. Balancing Fixed and Growth Mindset
12. Growth Mindset
13. Right Environment
14. Courage Building
15. Mindset Growth Through Personality Awareness
16. Components and Strategies for Growth Mindset
17. The Growth Mindset: Examples for Practical Mindset of Idealists
18. What is Intelligent Thinking?

19. Types of Thinking
20. Cognitive Thinking Pattern of Idealists
21. Integrative Thinking
22. Adaptive Intelligence Beyond Cognitive Agility
23. AI-Enabled Integrative Thinking
24. Why do some ideas fail?
25. Top Five Traits of Successful Startup Founders
26. Cognitive Functions Ti and Te
27. How to Develop Ti
28. How to Develop Te
29. Idealists in Business
30. Navigating Innovation Realities for Business Success
31. What is Excellence?
32. Personality Diversity for Team Success
33. How do you choose the right opportunity?
34. Cognitive Functions Se and Si
35. How to Develop Se
36. How to Develop Si

Elevate your adaptability and master the art of practical intelligence with this indispensable resource that offers tangible solutions and real-world applications.

Fearless Empathy Series

Book 1 : Set Smart Boundaries

"Want to find the answers to the questions holding you back? *Ask yourself these five questions:*

1. Are you tired of feeling like a pushover in your personal and professional relationships? It's time to take control and set clear boundaries in the workplace.
2. Are you fed up with constantly giving in to others' demands and not standing up for yourself? Let's work on developing assertiveness skills in your personal and professional life.
3. Do you need help communicating your needs and wants confidently and effectively in your personal and professional life? Let's explore ways to improve your assertiveness.

4. Are you feeling drained and unappreciated in your personal and professional relationships? It may be time to take a hard look at how you set and enforce your boundaries.
5. Are you ready to take charge of your life and start living in alignment with your values in your personal and professional life? Let's work on building your assertiveness and boundary-setting skills.

"Set Smart Boundaries: is a comprehensive guide for anyone looking to improve their relationships, advance their career, and achieve their goals. **This book provides a specific, measurable, achievable, realistic, and time-bound approach to setting boundaries.**

The natural ability to set boundaries is different for everyone. Certain people must consciously impose it as they cannot set boundaries naturally. In the MBTI 16Personality types, Intuitive & Sensory Feelers, require training in setting limits.

Get ready for an eye-opening adventure as this book takes you on a journey through the subtopics below, unravelling intriguing insights and captivating stories.

1. Why Spot Takers?
2. Definition of Boundaries
3. Who should set boundaries?
4. How to spot takers?
5. Toxic behaviours in people.
6. Why learn mindful giving?
7. Material Boundaries
8. Financial Boundaries
9. Givers & Takers
10. Why start valuing yourself?
11. Social Boundaries
12. Workplace boundaries
13. Religious and Intellectual Boundaries
14. Why should we protect our vibes?
15. Social Media Boundaries
16. How can we create social media boundaries?
17. How to build boundaries with mobile phones?
18. How to build boundaries with online meetings or classes?

19. How can we prevent abuse?
20. Personal Boundaries
21. Cyber Bullying
22. Sexual Boundaries
23. Why Stop being taken for granted?
24. Time-based Boundaries
25. Trauma Response
26. Signs of Poor Boundaries
27. Signs of being taken for granted
28. Traits prone to be taken advantage of
29. How can you stop people from taking advantage of you?
30. Exceptions to SMART Boundaries

This book is packed with practical advice, actionable tips, and real-life examples to help you set the boundaries you need to achieve success and happiness. Whether you're dealing with a demanding boss, a toxic friend, or a controlling partner, "Set Smart Boundaries" provides a step-by-step approach to help you take control of your life, career, and relationships.

Book 2 : Master Mindful No

Are you tired of feeling overwhelmed in a world that never stops demanding your attention?

Ask yourself these five questions:

1. Do you feel like you're constantly distracted and putting other people's needs ahead of your own, even if it means sacrificing your well-being? Let's Identify if you're a people pleaser and break free from this habit, prioritizing your needs for a fulfilling life.
2. Do you struggle with being true to yourself and practicing self-care? Let's discover practical ways to practice real self-care and be more authentic for a more fulfilling life.
3. Are your fears holding you back from achieving your goals and living your best life? Let's explore your concerns and move forward with confidence and purpose.
4. Do you struggle with managing guilt and difficulties when you say "no"? Let's strategize for managing

guilt and difficulties that may arise when speaking "no" to maintain healthy relationships and confidence.
5. Have you ever struggled with saying "no" without damaging your relationships or professional reputation? Let's Learn to say "no" positively and effectively, prioritizing our own needs while respecting the needs of others.

"Master Mindful No" offers practical strategies to help you filter distractions, overcome manipulation, and eliminate fear and guilt to succeed in a constantly demanding environment.

The natural ability to say No is different for everyone. Certain people must consciously learn it as they cannot be assertive naturally. In the MBTI 16Personality types, Intuitive & Sensory Feelers require training in prioritizing their needs.

This book takes you through the subtopics below, unraveling intriguing insights and captivating stories.

1. What is Mindfulness?
2. What is Mindful 'No'?
3. What is Distraction?
4. Why are we distracted?
5. Types of Distractions
6. Cost of Distraction
7. Practicing Mindful 'No' with Distractions
8. Root Causes of People Pleasing Behaviour
9. Courage Vs. Warmth
10. Manipulation Definition.
11. Signs of Manipulation
12. Practicing Mindful No with Manipulation.
13. What is Authenticity?
14. Authenticity and Sincerity
15. How to be Authentic?
16. Cost of Authenticity
17. Cons of authenticity at work.
18. Are your values limited?
19. Is fear what is standing in your way?
20. Why do we fake fear?
21. What are the common fake fears? How can we move forward?

22. Signs that you are living in fear
23. Mindfulness to transform fear.
24. Present-day fears of our life
25. Definition of Guilt
26. Shame Vs. Guilt
27. Shaming
28. Overcoming Guilt
29. Managing Guilt at Work
30. Guilt and Shame as Marketing Tools
31. Principles of Positive No
32. Power of Positive No in Negotiations
33. Saying No as a Productivity Hack
34. How to Say Positive No
35. Saying No at Work

With practical exercises, real-life examples, and thought-provoking insights, "Master Mindful No" is the ultimate resource <u>for anyone who wants to learn how to say "no" mindfully, with confidence and purpose.</u> Whether you're struggling with people-pleasing tendencies or feeling overwhelmed by commitments, this book will help you navigate the complexities of modern life and live a more fulfilling, peaceful life.

Book 3: Conquer Key Conflicts

"Do you crave to break free from the relentless cycle of adjustment?"

Ask yourself these five pivotal questions:
1. Are you tired of avoiding conflicts and arguments and ready to develop the courage to face them head-on? Assess your growth values.
2. Are you seeking practical strategies to transform conflicts into opportunities? Uncover opportunities for success.
3. Do you want to understand the benefits of conflicts and learn how to manage them effectively? Navigate for positive outcomes.
4. Are you ready to choose healthy battles and leave your comfort zone? Discover more authentic answers.
5. Do you want constructive confrontation? Foster a positive attitude and deepen relationships.

"*Conquer Key Conflicts*" offers 7 **Effective Strategies** to Stop Avoiding Arguments, Develop the Courage to Disagree, and Achieve Deserving Results in a Challenging Environment.

The natural ability to face conflicts is different for everyone. Certain people must consciously learn it as they cannot be assertive naturally. In the MBTI 16Personality types, Intuitive & Sensory Feelers require training in prioritizing their needs.

Discover a transformative guide to navigating conflicts with confidence and achieving excellent results. Explore the drawbacks of conflict avoidance, unlock the potential benefits of conflicts, and learn to choose healthy battles. This book takes you through the subtopics below, unraveling intriguing insights with examples.

1. Definition of Conflict
2. Triggers of Conflicts
3. Types of Conflict
4. Personality Types & Values
5. Values of MBTI Types
6. Personal Value Conflicts
7. What is Conflict Avoidance?
8. Signs of Conflict Avoidance
9. Conflict Avoidance or Value imbalance?
10. Values for growth
11. Result of Conflict Avoidance in Organisation.
12. Tips for Overcoming Conflict Avoidance
13. Should we encourage conflicts?
14. Disagreeing at work
15. Advantages of Conflicts at Work
16. Merits of Difficult Conversations
17. Conflict of Interest
18. Examples of Conflict of Interest at Work
19. Differentiating Conflicts
20. Arguments to Avoid
21. Choosing Value Conflicts for Success
22. Supporting the Right People in Conflicts
23. Conflicts and their Roots
24. Effective Confrontation
25. Mindful Confrontation
26. Tips for Constructive Confrontation.
27. Impact of Communication on Conflict Resolution

28. Effective Communication Strategies for Constructive Confrontation.
29. Importance of Active Listening
30. Applications & Benefits of Active Listening
31. Conflict Management Skills
32. Thomas-Kilmann Conflict Mode Instrument
33. Strategies for Value-based Conflict Resolution
34. Systems for Managing Workplace Conflicts
35. The Seven Strategies to Conquer Key Conflicts

From understanding the nature of disputes to **embracing healthy confrontation**, this book takes you on a journey of self-discovery and empowerment. *With practical strategies for resolution, you'll develop the courage to disagree and achieve positive outcomes in any challenging environment.*

Book 4: Build Emotional Resilience

Are you tired of being swept away by the chaos of life, losing your balance in the turbulence?
Ask yourself these five questions:

1. What if you could navigate life's challenges without being overwhelmed? Get ready to rewrite your relationship with adversity.
2. Have you ever felt your emotions spinning out of control? Dive into the heart of emotional imbalance and discover the tools to regain control.
3. What if you could break free from emotional dependence? Explore the empowering merits of emotional independence and learn how to cultivate it.
4. Can emotions indeed be your allies? Gain the power to make informed decisions and forge a more authentic path.
5. What if you could gracefully dance through life's ups and downs? Discover how to cultivate this invaluable skill and watch as life's challenges transform into opportunities for growth.
Step into emotional reinforcement, where you'll learn how to nurture and magnify the emotions that uplift you. This is your guide to mastering Emotional Resilience and thriving in chaos. *With captivating stories, practical exercises, and eye-opening insights, this book is your companion on the journey to a calmer, more empowered you.*

The innate capacity to process emotions varies among

individuals. Some people may need to consciously develop this skill, especially if they possess heightened sensitivity. According to the MBTI 16 Personality Types, individuals categorized as Intuitive and Sensory Feelers may benefit from acquiring Emotional Resilience through training.

This book takes you through the subtopics below, unraveling intriguing insights and captivating stories.

1. Emotional Resilience
2. Emotional Intelligence Vs. Emotional Resilience
3. Factors Influencing Emotional Resilience
4. Negative Emotions
5. Emotional Setbacks
6. Relevance of Emotional Resilience
7. People Vulnerable to Frequent Emotional Imbalance
8. Highly Sensitive Persons (HSPs) and Empaths
9. Emotional Imbalance and Energy
10. Emotional Imbalance Based on Personality Type Cognitive Functions
11. Emotional Imbalance Based on Trauma
12. Emotional Independence
13. Ways to Achieve Emotional Independence
14. The Power of Detachment
15. The Power of Non-Reaction
16. Emotions are built, not built-in
17. Three Ways to Better Understand Your Emotions
18. Premeditatio Malorum
19. The Theory of Constructed Emotions
20. Strategies for Emotional Intelligence at Work
21. The Science of Romantic Love
22. Three Methods to manage emotions in the workplace
23. Habits of Emotionally Disciplined Leaders
24. Emotional Agility
25. Radical Acceptance of Emotions
26. Measuring Emotional Agility and Resilience
27. Emotional Agility for Workplace Success
28. Emotional Agility for Effective Leadership
29. Shame Resilience Therapy
30. Resilience in the Face of Harsh Criticism
31. Energy Frequencies, Emotions, and Healing
32. 9 Strategies for Lifting Your Mood Immediately:

33. Tips for Naturally Boosting Energy Levels
34. 10 Ways to Enhance Emotional Resilience
35. The Healing Power of Connection
36. Holistic Approaches to Emotional Resilience and Well-being

This book unveils the intelligence hidden within your emotions and teaches you how to harness their wisdom. ***Emotional Resilience is your ticket to fluidly adapting to any situation.***

Book 5: Develop Vital Connections

Are you weary of navigating the ruthless battleground of modern life without a safety net?

Ask yourself these pivotal questions:
1. What if you could harness the power of effective communication and self-expression to overcome life's challenges confidently? Understand the profound impact of connections on your growth, happiness, and success.
2. Have you ever considered the advantages of a robust support system in our competitive world? Discover the benefits of nurturing personal and professional relationships.
3. Struggling to establish vital connections? Learn to identify and conquer common barriers that hold you back.
4. How can you choose connections that elevate your life? Gain the wisdom to cultivate relationships that truly empower you.
5. Need techniques for lasting connections? Equip yourself with practical strategies to build meaningful bonds.

Step into vital connections, where you'll learn the art of mastering effective communication, empowering your self-expression, and enhancing your value in this competitive arena. *This book isn't just a guide; it's your steadfast companion on the journey toward a more connected, thriving you.*
<u>Innate connection-making abilities differ among individuals. Some may need to develop this skill, especially if they prefer solitude consciously. According to MBTI's 16 Personality Types,</u>

<u>Introverted Intuitives can benefit from strengthening their connection-building skills.</u> In a world where the ability to build and maintain vital connections is your golden ticket to success, whether you're a natural social butterfly or someone who could use a bit of extra guidance, **"Develop Vital Connections" reveals the intelligence concealed within the craft of connection-building, teaching you how to harness its incredible potential.**

This book will take you on an exhilarating journey through the following key topics, unraveling intriguing insights and sharing captivating stories:
1. Vital Connections
2. Attachment Styles
3. Factors of Connections for Growth
4. Effective Communication vs. Self Expression
5. The Power of Networks
6. Connecting with a Common Story
7. Connections for Opportunities and Job Advancement
8. Connections to Enhance Learning and Knowledge Sharing
9. Amplifying Influence Through Meaningful Connections
10. Connections to Console and Navigate Challenges or Distress Times
11. Connections to Fulfil Life
12. Introverts and Extroverts
13. Why do Introverts Avoid Small Talk?
14. Why Do Some People Avoid Socializing?
15. How Trust Issues Impact Communication
16. The Connectedness Corrective
17. Inability to Identify the Value of Communication
18. Chances for Establishing Connections
19. Building Meaningful Connections for Your IKIGAI
20. Connections for Adapting to Change
21. Balancing Patience and Proactivity
22. What is the reason behind our innate drive for connection?
23. Knowing Personality Types for Connection
24. What Is Effective Communication?
25. Mastering Effective Communication
26. Communication Tips for Maximum Impact

27. The Power of 'Because' to Influence Behavior
28. Gesticulation and Nonverbal Cues in Effective Communication
29. How to Make People Feel Heard?
30. Building Lasting Connections
31. Authentic Connections
32. 10 Effective Ways to Build Solid Professional Connections
33. Three Predictors of Lasting Connections

Join us on this transformative adventure and witness your life evolve into a tapestry woven with flourishing connections, boundless opportunities, and unwavering support.

Clear Career Inclusive Series

Book 1: Raising Your Rare Personality

Find who you are to be your best!

What is your personality type? Are you the right fit for your career? Who is a rare personality type? This book provides all the answers. Psychology is the scientific study of mind and behavior. Understand how psychology defines your unique type, growth potential, and suitable careers. Myers-Briggs Type Indicator (MBTI), a tool to identify personality typology, classifies people into 16Personalities. You can belong to any one of these 16 personality types based on your psychological preferences. Some personality types are stated as rare personality types as per MBTI. The personality type INFJ has been explored in-depth in this book. The purpose of this book is to show solidarity to who you are, identify suitable careers for all MBTI types, with a focus on the rare personality types.

Key Learnings from the book - Raising Your Rare Personality

Chapter 1 MBTI Personality Types
1. What are MBTI Personality Types?
2. How can you understand your Personality Type?
3. What are the 16Personalities?
4. Who are Rare Personality Types?
5. Who is the Rarest Personality Type?

Chapter 2 MBTI Cognitive Functions
1. What are Cognitive Functions?

2. What are the 8 Cognitive Functions?
3. What is a Primary Cognitive Function?
4. What is a Shadow Cognitive Function?
5. Cognitive Functions of all MBTI Personality Types

Chapter 3 INFJ Primary Cognitive Experiences
1. What are the Primary Cognitive Functions of an INFJ?
2. How does Introverted Intuition behave?
3. How does Extraverted Feeling behave?
4. How does Introverted Thinking behave?
5. How does Extraverted Sensing behave?

Chapter 4 INFJ Shadow Cognitive Experiences
1. What are the Shadow Cognitive functions of an INFJ?
2. How does Extroverted Intuition behave?
3. How does Introverted Feeling behave?
4. How does Extroverted Thinking behave?
5. How does Introverted Sensing behave?

Chapter 5 Rare Personality Types and Growth
1. Growth potential Function of MBTI Personality Types.
2. What are Functional Pairs?
3. How Intuition works in Rare Personality Types?
4. Strength and Weakness of INFJ Personality Types
5. Famous Personalities of all MBTI Rare Personality Types

Chapter 6 Careers for your Personality
1. Functional Pair strength for all Personality Types
2. Careers for Intuitive Feelers
3. Careers for Intuitive Thinkers
4. Careers for Sensory Feelers
5. Careers for Sensory Thinkers

Resources
Free Test links for finding MBTI Personality, Enneagram, Socionics, Big 5, DISC, Holland Code Job Aptitude Test, etc. are included in the book.
"A man's true delight is to do the things he was made for." – Marcus Aurelius

✓ **Find Yours!**

Book 2: Upgrade as Futuristic Empaths

Find your strength to give your best!

Are you an empath? Do you know what an empathy trap is? How can you transform empathy into a strength and build successful careers?

Empaths have intuitive feelings (owing to the cognitive functional pair "NF" in their personality type) as their psychological preference. Personality types ENFP, ENFJ, INFJ, and INFPs are natural empaths as per the **MBTI Personality types** according to www.16personalities.com and www.Truity.com. Empaths are also called **Idealists & Diplomats. Highly Sensitive People** belong to these MBTI types. To face the realities of the world and to be successful in endeavours which have larger impacts, empaths need to embrace practicality and rise above their personality stereotype or one-sidedness.

Dr.Dario Nardi, Author of the book **Neuroscience of Personality**, suggests transcendence or the individuation process, a term coined by **Carl Jung,** the essence of which is to have an integrated personality growth. Empaths have a larger role to play in this world and most of them are underplaying their natural strength.

By adopting the 5 key steps discussed in this book, anyone, especially empaths can easily find their career paths to success, thereby leaving a positive impact on this world.

Key Learnings from the book - Upgrading as Futuristic Empaths.

Chapter 1 Understanding Empaths

1. Empathic People or Empaths
2. Empathy Dilemma
3. The Value of Empathy
4. Practising Empathy
5. The Empathy Trap
6. Use of Empathy in day-to-day life
7. Empathy and Business
8. Empathy and Leadership

Chapter 2 Finding your Strength

1. Empath's Strength, Weakness & Dilemma
2. Empaths as Employees
3. Clifton Strengths
4. Machiavelli's Dilemma
5. Empath's Choice
6. Empathy as a strength in daily life
7. Fearless Empathy
8. Nurturing Empathy

Chapter 3 Developing Your Profile

1. An Empath's Growth Cognitive Function
2. Moving from One-sidedness to individuation
3. Challenges of One-sidedness for Empaths
4. The Magic Diamond for Integrated/Transcendent Judgement & Perception
5. Preferred Growth of Empaths Cognitive Functions
6. The Spiral Development of Cognitive Functions
7. Using Empathy as a Strength
8. Essentials for Building an Empath's Profile
9. Careers and Majors for Empaths

Chapter 4 Finding Your Market Niche

1. Sustainable Development Goals in Business
2. Future Job Skills
3. Selecting a Career for Empaths
4. Challenges of Workplace Toxicity
5. Future of Jobs for Empaths
6. Empaths and the Gig Economy

Chapter 5 Connecting & Networking

1. The Power of Social Connection
2. Why are we not Connecting?
3. Impact of Networking
4. Managing Digital Distraction

Chapter 6 Creating Opportunities

1. Opportunities for Empathy in Business
2. Opportunities in Sustainability
3. Empathy Revolution

"Objective judgment, now, at this very moment. Unselfish action, now, at this very moment. Willing acceptance — now, at this very moment — of all external events. That's all you need." - Marcus Aurelius

✓ **Find How!**

Book 3: Onboard as Inclusive Leaders

Find Your Potential to Impact the Best!

How Inclusive are you? Are you unconsciously biased?

Do you promote Psychological Safety?

This book will help you find answers and enable you <u>Onboard as Inclusive Leaders.</u>

Innovation, financial performance and employee productivity are indispensable for business growth. Inclusion helps in achieving these objectives of business. Diversity in line with inclusion and equity creates a sense of belonging in employees.

This book helps to develop the essential qualities required to be hired as an inclusive leader; **understand unconscious biases, the importance of psychological safety and how it has an impact on workplace productivity.**

The book also gives you the free test links to understand your MBTI personality type, strength, and Bias Tests (The Implicit Association test - Harvard University)

Key Learnings from the book – Onboard as Inclusive Leaders

Chapter 1 Knowing Inclusion

1. Why do we need Inclusive Leaders?
2. What is an Inclusive Workplace?
3. Features of an Inclusive Workplace
4. Challenges of Inclusive Workplace

5. Merit based Inclusion
6. Who is an inclusive leader?

Chapter 2 Inclusion Gap

1. Facts of Diversity & Inclusion
2. Microaggression
3. Unconscious Bias
4. 16 Unconscious Biases
5. Bias Test (The Implicit Association Test)
6. The Cost of Unconscious Bias

Chapter 3 Inclusion in Practice

1. Inclusion in the workplace
2. Inclusion Strategies at Ingersoll Rand
3. Inclusion Mandate
4. Expectations of Gen Z
5. Disability Inclusion
6. LGBTQ+ Inclusion
7. Six Signature Traits of Inclusive Leaders
8. Risks of Casual Diversity Programs

Chapter 4 Inclusion Participants

1. Types of Inclusion
2. Physical Inclusion
3. Psychological Inclusion
4. Importance of Assertiveness for Empaths at work
5. Empathy and Neuroscience of Personality Types
6. Preparing the Team for Inclusion

Chapter 5 Inclusion Process

1. Inclusion Strategy
2. Psychological safety
3. International Standards for Inclusion Process
4. Inclusive Job Posting
5. Inclusive Hiring
6. DEI Interview Questions
7. Disparate Treatment & Disparate Impact

Chapter 6 Inclusion Measurement

1. Measurement of Inclusion
2. Gartner Inclusion matrix
3. How Inclusive is your leadership?
4. Fundamental Interpersonal Relations Orientation (FIRO®)
5. Empathy & Inclusion Measurement
6. Industry Measurement of Diversity & Inclusion

"If someone can prove me wrong and show me my mistake in any thought or action, I shall gladly change. I seek the truth, which never harmed anyone: the harm is to persist in one's own self-deception and ignorance."
— Marcus Aurelius

We need more inclusive leaders who will consider others in their decisions and that alone can give rise to sustainable development and positive impacts for people and the planet.

✓Find How

Acknowledgement

My gratitude to the readers of my book, for your time and reviews, and to all my well-wishers for your support. I am indebted to all who reached out to me with feedback and input. I have to start by thanking my family, friends, and classmates for their encouragement, counsel, and good-natured jibes. Extending my wholehearted gratitude to everyone on the Author Freedom Hub, special thanks to Som Bathla for his vote of confidence and my fellow authors for their unbounded support. To Anita Jocelyn for her editorial help towards the completion of my book. I am grateful to Mr. Sareej for his efforts towards the beautiful cover design. I thank my friends and colleagues who helped me with their insights and experiences of their work place inclusion. Your inputs were critical in the completion of this book and helped me gather information to cover this topic in details for my readers. In no way at all the least, I am very thankful to my spouse Jo and our son Yakob for helping me out immensely by allowing me space and time to pursue my interests and creating a conducive environment to achieve my goals. To my mother Prof. Thresiamma Sunny, I am thankful for her unwavering support and inspiration to always deliver my best.

I could not have done it without you all.

References

Chapter 1

1. Gerald Ratner - Wikipedia
2. Moneyball (film) - Wikipedia
3. 12 Analytical Intelligence Examples (2023) (helpfulprofessor.com)
4. What Is Analytical Intelligence and How Is It Used? (With Examples) | TopResume
5. What Is Analytical Intelligence and How Is It Used? | Indeed.com

Chapter 2

1. How Amazon Survived the Dot-Com Bubble | HBS Online
2. Dot-com bubble - Wikipedia
3. Of 2 Minds: How Fast and Slow Thinking Shape Perception and Choice [Excerpt] | Scientific American
4. Sales Psychology | Psychology of Sales | Sales Psychology Tips (sloovi.com)
5. How to Make Great Decisions, Quickly (hbr.org)

Chapter 3

1. Bay of Pigs Invasion - Wikipedia
2. Bay of Pigs - Groupthink (globalsecurity.org)
3. Groupthink: Definition, Signs, Examples, and How to Avoid It (verywellmind.com)
4. Cognitive bias - Wikipedia
5. Cognitive Bias List: Common Types of Bias (verywellmind.com)
6. Beware the Dangers of Cognitive Bias - National Commission on Correctional Health Care (ncchc.org)
7. Cognitive Bias: Understanding How It Affects Your Decisions (healthline.com)
8. Critical thinking - Wikipedia
9. Why Your Business Needs Critical Thinking (forbes.com)
10. 3 Simple Habits to Improve Your Critical Thinking (hbr.org)

11. J. Clifford, Magdalen M Boufal and J. E. Kurtz. "Personality Traits and Critical Thinking Skills in College Students." *Assessment*, 11 (2004): 169 - 176.
12. https://doi.org/10.1177/1073191104263250.

Chapter 4

1. Toyota Production System | Vision & Philosophy | Company | Toyota Motor Corporation Official Global Website
2. The Benefits of Being a Slow Thinker | Psychology Today
3. Cognitive reflection test - Wikipedia
4. Gone Baby Gone - Wikipedia
5. Logical reasoning - Wikipedia
6. https://curiousdesire.com/why-logic-is-important/
7. 20 Behaviors That Reveal Someone Is a Logical Thinker (powerofpositivity.com)

Chapter 5

1. Review, Harvard Business. HBR Guide to Critical Thinking (p. 47). Harvard Business Review Press. Kindle Edition.
2. The Five Pitfalls Of Problem-Solving - And How To Avoid Them (forbes.com)
3. The strengths and weaknesses of every personality type | World Economic Forum (weforum.org)
4. Know Your Myers-Briggs Type Indicator? Use Your Results To Your Advantage (forbes.com)
5. 14 Effective Problem-Solving Strategies | Indeed.com

Chapter 6

1. 93 Billion Hours Of Netflix: What The Data Reveals (forbes.com)
2. Using Big Data And Data Analytics For Better Business Decisions (forbes.com)
3. The Advantages of Data-Driven Decision-Making | HBS Online
4. How to Improve Your Analytical Skills | HBS Online
5. Using Data To Make Better Decisions (forbes.com)
6. 10 Best AI Tools for Decision-Making That Teams Love | ClickUp

7. Frameworks for Intelligent Decision-Making | Psychology Today
8. Harvard Business Review; Review, Harvard Business. HBR Guide to Making Better Decisions (p. 176). Harvard Business Review Press. Kindle Edition.

Copyright © 2023 by Devi C.Sunny

All Rights Reserved. No part of this book may be reproduced or used in any manner without the written permission of the copyright owner except for the use of quotations in a book review.

Printed in Great Britain
by Amazon